Industrial Engineering

Industrial Engineering Techniques for improving operations

Edited by
Vas Prabhu
Principal Lecturer in Production and Operations Management
Newcastle-upon-Tyne Polytechnic

and

Malcolm Baker
International Management Consultant and Director of Training
Institute of Production Control

McGRAW-HILL Book Company (UK) Limited

London · New York · St Louis · San Francisco · Auckland · Bogotá
Guatemala · Hamburg · Johannesburg · Lisbon · Madrid · Mexico
Montreal · New Delhi · Panama · Paris · San Juan · São Paulo
Singapore · Sydney · Tokyo · Toronto

Published by
McGRAW-HILL Book Company (UK) Limited
MAIDENHEAD · BERKSHIRE · ENGLAND

On behalf of the
Institute of Production Control
STRATFORD-UPON-AVON · WARWICKS · ENGLAND

British Library Cataloguing in Publication Data

Industrial engineering: techniques for improving operations.
  1. Industrial engineering
  I. Prabhu, Vas II. Baker, Malcolm
  658.5   T56
  ISBN 0-07-084932-3

Library of Congress Cataloging-in-Publication Data

Prabhu, Vas
  Industrial engineering.

  Includes index.
  1. Industrial engineering. I. Baker, Malcolm.
  II. Title.
  T56.P698   1986   658.5   86-10273
  ISBN 0-07-084932-3

LR 02168

Copyright © 1986 Institute of Production Control. All rights reserved. No part of this publication may be reproduced, stored in a retrieval system, or transmitted, in any form or by any means, electronic, mechanical, photocopying, recording and/or otherwise, without the prior written permission of the Institute of Production Control. This book may not be lent, resold, hired out or otherwise disposed of by any way of trade in any form of binding or cover other than that in which it is published, without the prior consent of the Institute of Production Control.

12345 CUP 89876
Typeset by Advanced Filmsetters (Glasgow) Ltd and
printed and bound in Great Britain at The University Press, Cambridge

Contents

Preface vii

Part 1 INDUSTRIAL ENGINEERING PRINCIPLES
1 Industrial engineering—the key 3
2 Introduction to work study 10
3 Historical development 16
4 What is management? 21
5 The basis for success—the human factor 27

Part 2 METHODS ENGINEERING
6 Methods engineering 33
7 Principles of methods engineering 40
8 Automation 52

Part 3 WORK MEASUREMENT
9 Work measurement 57
10 Computerized systems of estimating, planning and standard data 75

Part 4 INTRODUCTION TO STATISTICAL METHOD AND OPERATIONS RESEARCH
11 Introduction to statistical method 95
12 Descriptive statistics 97
13 The normal distribution 114
14 Estimation and confidence intervals 121
15 Operational research 128

Part 5 INDUSTRIAL ENGINEERING ON THE SHOP FLOOR
16 Shop floor data collection 139
17 Production scheduling of multistage and jobbing production units 148
Appendix Summary of some common terms used 166
Index 169

Preface

Several major research studies of manufacturing industry in the United Kingdom done during the 1970s came to one key conclusion, namely that there was still considerable scope for improvement in industry's resource productivity levels as well as in its ability to provide a high customer service level. Shop floor production efficiencies of no more than 30 per cent and poor delivery records were among some of the findings of these studies. Not surprisingly, other national studies conducted in the early 1980s have shown a considerable lack of use of modern management techniques by production management personnel. Even the traditional and well-established techniques of 'method study' and 'work measurement' were being used in only about 35 per cent of companies.

Also, during the last decade several external factors have placed additional pressures on manufacturing industry's capability of performing effectively. For example, competition, especially from overseas markets, is becoming more severe and the customer's demand for cheaper and better-quality products, coupled with shorter delivery times, is placing considerable strain on the efficiency of its manufacturing function. The ability to implement and (more importantly) manage the new technologies of robotics, computer-aided design and manufacture, into industry will also have a significant bearing on its performance capability. Given this general backcloth, the skills and techniques of industrial engineering therefore have an important role to play in any revitalization process.

This book, which is aimed primarily at the practising manager, provides an overview of some of the main subject areas of industrial engineering as well as an insight into some of the more recent applications of computers to them. It is not intended to be a basic training book; indeed some of the more traditional techniques of method study and work measurement have not been covered in depth as this is already done by several books currently on the market. The contents should be read in conjunction with other, basic texts on the subject, as this book will provide the practising manager with an updated view of certain aspects. It should therefore be of particular interest to all

PREFACE

managers at whatever level and within any functional area who are responsible for managing some part of the company's resources, especially production managers and works managers.

The contents of this text are essentially drawn from the course notes that the Institute of Production Control has developed and used successfully over several years on a whole range of its short management courses and seminars for senior and middle managers from manufacturing industry, not only in the Western developed world but also in the Eastern Bloc and in developing nations. The Institute felt that this material, though originally developed for its own use, would be of considerable interest and potential use to the vast majority of managers who typically cannot attend external daytime courses. The aim is to reach as many industrial managers as possible through this publication.

The material is divided into five separate parts. Part 1 reinforces the principles of industrial engineering and especially the importance of the human element for its successful application in industry. Part 2 surveys a range of techniques in methods engineering, including the impact of automation on manufacturing processes. In Part 3, on work measurement, the basic techniques are briefly reviewed and the third-generation techniques of predetermined motion time systems (MOST—Maynard Operation Sequence Technique) are explained. Also, details of several computerized systems of estimating, planning and standard data are described. Part 4 deals with the basic statistical and operational research concepts and techniques that an industrial engineer would find useful in practice. Finally, Part 5 looks at the role of industrial engineering on the shop floor; in particular it describes shop floor data collection methods (both manual and computer systems) as well as computerized scheduling systems.

Either the book can be read straight through from beginning to end, especially by the younger reader or the novice, in order to gain an overview of the major aspects of this area of study, or it can be read selectively, probably by the more experienced practitioner who may wish to increase his or her knowledge and understanding in either the techniques or the computing side of the subject area, referring only to those sections that he or she wishes.

Open learning is a highly cost-effective way of learning, allowing people to select a time, a place and a pace to suit their individual needs. This book will provide a valuable information resource in its own right to both companies and individuals involved with open learning projects. However, one point every reader should bear in mind when reading it is the need to discuss the topics within it with colleagues at work or other interested persons, as so often the learning process is enhanced by trying to apply or even relate the theoretical concepts and principles of these new techniques to one's own working environment and experience. In this respect, the Institute of Production Control would be willing, if required, to provide any additional information on an individual basis or through participation in its own training programmes.

K. Roberts
General Secretary

Discrimination between the sexes
In a book on this subject it has not been possible to eliminate the use of gender and still retain a fluent text. A deliberate distinction between the sexes is not implied.

Acknowledgements

We would like to thank H. B. Maynard & Co. Ltd for permission to reproduce certain tables and data from the *MOST Practitioners Training Manual*. We would also like to thank Westborough Computer Services Limited, Methods Workshop Limited and Kewill Systems plc for permission to reproduce some of their material.

Part 1

Industrial engineering principles

1

Industrial engineering—the key

1.1 Effectiveness in operations

An operating system is a configuration of resources combined for the provision of goods and services. The physical resources are usually the four Ms: materials, machinery, manpower and money which form the inputs that are converted to the outputs required by the customer. Aircraft, rail or bus services, hotels and superstores, mining extraction and refining, public services, hospitals and parts, product or process manufacture—all are operating systems; operations management is concerned with providing such systems with effectiveness and efficiency in terms of quality, productivity, investment and human development.

Do we perhaps pay too much attention to what should be done and to so-called financial controls and issues to the neglect of operations? After all, the public are often directly aware of shortcomings in operations and services, e.g., trains not running on time, no buffet service, poor food, etc.

HOW CAN WE ACHIEVE EFFECTIVENESS IN OPERATIONS?

Generally, there are two main ways, as follows:

1. By training and appointing competent and professional operations managers.
2. By using the professional services of those trained to make operations more effective, namely industrial engineers.

1.2 What are industrial engineers?

They are people skilled in applying scientific analysis, technical design, management techniques, financial appraisal and human relations principles in order to improve quality, productivity, investment and human development in operations. Industrial engineers are not line managers, but constitute, rather, a support group responsible for managing change. They are *not* operations managers.

Effectiveness in operations demands that productivity of all resources is at a high level, particularly when the product is competing within the world market-place. Perhaps the key role in increasing productivity lies with the function of industrial engineering; this the American Institute of Industrial Engineering has defined as being:

> concerned with the design, improvement and installation of integrated systems of men, materials and equipment. It draws upon specialized knowledge and skill in mathematical, physical and social sciences together with the principles of engineering analysis and design to specify, predict and evaluate the result to be obtained from such systems.

Others have redefined industrial engineering as

> the profession which has the prime objective of making companies more competitive.

What is industrial engineering? Perhaps a list of common disciplines would be useful to start with:

1. The layout of factories and design of work flows.
2. Operation methods analysis and design.
3. Work measurement systems.
4. Production and materials control systems.
5. Computer-aided design (CAD), computer-aided manufacture (CAM).
6. Robotics.
7. Production engineering (low-cost automation (LCA), machines and equipment, etc.)
8. Cost evaluation and control.

What makes an industrial engineer different from other management specialists? Mainly the following:

1. Having the necessary theoretical training in the disciplines involved.
2. Having experience in applying these disciplines in real-life working situations.
3. Being trained to implement successfully.
4. Having the necessary time to spend on the project in hand; the manager is too involved in the day-to-day running of the function for which he is responsible.
5. Keeping up to date and being a member of a professional institute promoting industrial engineering.

Industrial engineering is based primarily on advances in knowledge which are, it has been concluded, 'the biggest and most basic reasons for the persistent long-term growth of output per unit of input'. The term 'advances in knowledge' is a comprehensive one: it includes both what is usually defined as 'technical knowledge', concerning the physical properties of things and how to make, process or use them in a physical sense, and 'managerial knowledge', the knowledge of business organization and of managerial techniques considered in the broadest sense. Such knowledge originates both in the United Kingdom and abroad, and is obtained in various ways: by organized research, by individual research and by simple observation and experience.

1.3 Productivity

Productivity measurement has been studied over many years. Most people recognize that the generation of wealth by individual companies, and indeed by nations, depends upon high levels of productivity. In simple terms this means getting the best use of our resources and the following ratio provides a measure of productivity:

$$\text{Productivity} = \frac{\text{output}}{\text{input}}$$

The term productivity, however, can be applied at different levels, depending upon what one is looking at. This is what makes productivity measurement difficult. At national level, for instance, one often sees indices of GNP (gross national product) per capita or per number of people in employment. Within the company it is normally the productivity of labour which is measured.

There are various measurements we can choose for the productivity of capital; perhaps the most usual is 'profit return on net assets employed'. The productivity of materials could be considered to be given by the following formula:

$$\text{Materials productivity} = \text{utilization index} \times \text{yield} \times \text{scrap index}$$

Perhaps the biggest problem in a company is to measure overall productivity. Some have suggested the value-added approach, where

$$\text{Overall productivity} = \frac{\text{net sales invoiced} - \text{purchased content of sales}}{\text{cost of wages and salaries}}$$

This certainly is one measure of productivity which a few companies use.

H. B. Maynard, a company internationally acclaimed for providing precise measurements of productivity, because of its work measurement system, has defined labour productivity as a combination of three factors, as follows:

$$\text{Labour productivity} = \text{methods index} \times \text{utilization} \times \text{operator performance}$$

INDUSTRIAL ENGINEERING

METHODS INDEX

This is the ratio of the work content of the current method being used, divided by the work content of the current best method for the industry concerned. It is not an easy index to define, but it certainly has a big effect on labour productivity. If, by method changes, the work content can be reduced, say, from eight minutes per piece produced to six minutes per piece produced, but has not yet been installed, then the methods index is

$$\frac{6}{8} \times 100 = 75$$

The basis for the ratio will change over time: someone else, for instance, may look at the same job and reduce the time per piece to three minutes, through a more automated process.

UTILIZATION

This is defined by the following formula:

$$\text{Utilization} = \frac{\text{attendance time} - \text{lost time}}{\text{attendance time}}$$

which is the amount of time worked divided by the time available to work. Generally, this is influenced by manufacturing support service and management supervision. It can be improved by:

- Having less breakdowns of equipment (maintenance).
- Planning work more thoroughly (production planning).
- Having less materials shortages (materials management).
- Organizing people for effective working (production management and supervision).
- Having less scrap and rectification (quality assurance and control).

PERFORMANCE

The performance of any person or machine can be achieved only by 'measuring' the work involved, normally in standard minutes (SM) or standards hours (SH) of expected work; the performance measure is expressed as follows:

$$\frac{\text{output (SH or SM)}}{\text{time spent on measured work (hours and minutes)}} \times 100$$

INDUSTRIAL ENGINEERING—THE KEY

PRODUCTIVITY CENTRES

The last decade has seen a rapid development of productivity centres throughout the world; attention has been focused on three main aspects of productivity:

1. Productivity is frequently used to describe the health of an economic unit, usually of a country. A large number of studies have been made *comparing* various economic and statistical indices which purport to indicate something about productivity.
2. Productivity improvement is often recognized as a way to stimulate future business success and therefore the wealth of a nation.
3. Studies have been made of how to measure productivity in its many aspects, while how to improve this productivity at micro level has been given a great deal of attention.

The School of Business Administration at the University of Western Ontario in Canada has made a survey of productivity institutions and published their names and addresses in a booklet entitled *Western Productivity Study*. 'Western' is from the name of the university and not because the study is limited to the Western nations; centres in the Eastern bloc and the Asian and developing nations are also included, so it is a world study.

As far as the United Kingdom is concerned, there is no national centre for productivity, although there are a few institutions concerned with aspects of productivity.

The *Western Productivity Study* provides the following numbers of institutions, not all of which are productivity centres:

Africa	17	Middle East	7
Asia	28	North America	132
Australia and New Zealand	6	United Kingdom	13
Western Europe (continent)	57	(no productivity centres)	
Eastern Europe	18		

Productivity is recognized as a key to future success in the United States, and so there are a great number of centres, with various names such as:

- Productivity centre
- Quality of working life centre
- Productivity institute
- Quality of working life program
- Office of productivity programs

Many of the developing nations have national productivity boards and centres, often set up with the aid of the United Nations Economic

INDUSTRIAL ENGINEERING

Development Organization, and these are an integrated part of their industrial life. They usually work in the following main areas:

1. Promoting the idea of productivity with the public and so changing attitudes, through public relations exercises, advertising, slogans, public displays, broadcasts, etc.
2. Providing research activities into areas for productivity measurement and improvement.
3. Promoting courses, lectures, seminars, workshops and conferences to provide knowledge of productivity subjects to delegates.
4. Undertaking advisory studies and implementation programmes on productivity improvement.

At present the main areas of activity which influence productivity are the following, and should be emphasized:

1. *Economic studies* At macro and micro levels, including managerial economics, econometrics and international aspects.
2. *Managerial studies* The management of organizations, the practice of management, management techniques, behavioural sciences, marketing and selling.
3. *Technological studies* The impact of computers and high technology in society, interaction, R and D, product design, value analysis and engineering, factory automation, and office automation.
4. *Industrial engineering*
5. *Industrial relations*
6. *Cultural aspects of productivity* The whole background of philosophy, religion, culture and science and its impact on the success of ratios, including what is called an organization's own 'culture'.

INCREASING PRODUCTIVITY

In many parts of the world there will be many significant changes as robots and computerized manufacturing systems replace significant numbers of semiskilled machine operators and unskilled labourers over the next three decades or so. Computer-aided manufacture and robotics will reduce the benefit of mass production relative to batch production and thereby reduce the existing pressures towards product standardization and against diversity and change.

Mass production, by its nature, is compatible only with stabilized or standarized products. Once mass production facilities are in play which have usually cost many hundreds of thousands or millions of pounds, their existence is a deterrent to innovation. Typically, future systems will mean output in smaller batches, with the possibility of incorporating more variation in product design (but with proper standards of tolerance, specifications and fixing devices) more easily into a product's character. However, there is still going to be a need for people in many areas of activity, and getting improvements from existing resources will still be of paramount importance. It is significant that this can be achieved with little capital expenditure through the technique of work study, which is the topic of the next chapter.

2

Introduction to work study

2.1 History and scope

Most people are good at studying work! Whenever a gang of men dig up the road or a mobile crane is used to lift goods to another position, large groups of people will normally gather: there is an inborn fascination about other people working. However, this is not work study in the sense we use the term in this book: a systematic and analytical procedure which has a method of going about its task.

Work study is not confined just to ways of studying work; its purpose is to make improvements to existing or new work situations. Indeed, many organizations have gained considerable benefits through using its specialists and its techniques.

Work study can be applied to any kind of work. It started as a scientific procedure at about the beginning of this century and was applied mainly to manual work, for instance shovelling materials. It was quickly applied to industrial processes where repetitive manual work rates were increased so that output per person improved. It was then applied to non-repetitive work, such as maintenance and road and rail haulage, and to indirect work, where one person services another, such as work handling and work planning. Also included were administrative clerical work, such as typing, filing, photocopying and collating, and public services work, such as refuse collection, street cleaning and maintaining public parks. Work study has now been applied to the design and layout of hospitals, particularly in the United States, and to the organization of airports, railway stations and bus terminals. There are now standards even for the operations of robot movements. So work study has universal significance. It certainly is a technique which can improve the efficiency of operations and maintain or improve our standard of living. It is particularly useful because it can be used to make improvements with very little capital expenditure.

These are a few criteria which are important, however:

1. Work study needs to be done by people who have received training in its

techniques. This does not imply that everyone should be trained to diploma level in work study and management services.
2. It is not necessarily true that a few people should become the only specialists in work study. The technique would be applied better if it were done as a group exercise, as with value analysis and quality circles. Perhaps one of the *worst* uses of work study is where one person superimposes his own will on another person or group of persons with very little discussion or involvement.
3. Work study practitioners who are fully trained in the techniques, for example who have studied for the Diploma of the Institute of Management Services and may be full-time members of the Institute, are needed to play a different role today: they should be the 'facilitators' of work study, to use a quality circle term, the ones who assist others to apply work study to their own jobs.
4. Like all management techniques and methodologies, work study should be initiated by top management, because it is a vital contributor to the future prosperity and growth of the organization. Unless those in the higher management team are directly aware of the advantages, implications and techniques of work study, take an active interest and give their full backing, the procedure will get nowhere.

2.2 Why do we need to use work study?

The main reasons are the shortcomings we can all see at our place of work: poor quality, delays, backlogs and inefficiency; we see both the effects of these upon what we ourselves do and also how they affect us personally, providing frustration, pressure, and perhaps anger when the situation is seemingly impossible. Each person has a unique personal responsibility for ensuring that such problems are overcome. The positive contribution which trades unions could make by ensuring that their members are part of a success story, which they themselves share, would be welcome indeed.

There is a vital need for the information that we look at and present to be factual and based on the right motive for the task in question; otherwise we revert to the 'them-and-us', 'management-and-worker' approach which has prevailed in this country. We are all workers, working for common aims, namely:

1. The prosperity and success of our organization, which in turn leads to more secure work for every person involved.
2. That everyone should share in the success, both in terms of working conditions and in emoluments (wages or salaries). Work study can and should be used to improve these.

3. The elimination of all the forms of frustration and stress which affect us at work caused by poor work organization. Personal conflict will not be eliminated so easily, but it certainly is reduced if people are encouraged to work together as a team rather than as individuals.
4. Making quality paramount, along with cleanliness and tidiness; this is the personal responsibility of all together.

Once we get things together we become more competitive, business expands and we need more people; jobs are created.

In a properly organized country there is no need for unemployment. There is always plenty of work to be done (*always*). Where there *is* unemployment, what is missing is the right way for work to be organized and paid for, and the motivation to enable people to gain the respect they most desperately need. One of the worst problems a man can have is to be 'out of work' so that he cannot properly support his family; a man also needs to have an interest in something to give him self-respect. Some people 'don't want to work' or 'won't work' because the rewards are not enough or they 'don't want to get involved', etc. It should be recognized by all that such people are living at 'sub-species' level. They are either too selfish or they have been 'drummed into the ground' by the conditions around them. In the first instance they should not receive benefits; in the second instance they should be helped, perhaps by working in community groups with others sharing the same problems.

2.3 Work study definitions

It is now time to define some terms:

1. *Work study* Those techniques, particularly *method study* and *work measurement*, used in the examination of human work in all its contexts and which lead systematically to the investigation of all the factors which affect the efficiency and economy of the situation being reviewed, with the aim of bringing about improvement.
2. *Method study* The systematic recording and critical examination of existing and proposed ways of doing work, as a means of developing and applying easier and more effective methods and reducing costs.
3. *Work measurement* The application of techniques designed to establish the time for a qualified worker to carry out a specified job at a defined level of performance.

Although much has been written about work study there are many variations in the detailed practice of its components. It is important, however, to be precise about its method of operation, because all too often the results expected do not materialize unless the approach is objective, methodical and practical.

INTRODUCTION TO WORK STUDY

Within the context of modern work study, there are recent developments in various techniques which are necessary for the specialist practitioner, which, with intensive training and experience, can make a real impact on effective working within any organization. These techniques are outlined later.

The procedure for work study follows a defined sequence:

- *Select* the work or job to be studied.
- *Record* the work through charting and measurement techniques.
- *Examine* the results of the analysis, e.g., critical examination.
- *Develop* new improved methods.
- *Install*: implement these.
- *Maintain*: monitor, and review the new methods to ensure that they are working properly.

2.4 Method study

The object of method study is to develop improvements in layout and methods in order to achieve minimum work content for the operations under examination. Typical of such improvements are:

1. Proper batching of work to avoid excessive changeover time.
2. Organization of materials storage and handling and of work flow.
3. Organization of plant and workplace layout to eliminate unnecessary walking and movement.
4. Development of methods which enable both hands to be used productively.
5. Development of methods for performing machine operations on more than one process at a time.
6. Development of productivity aids, such as low-cost automation.
7. Design for production so that cost is reduced or eliminated.

The investigative work for method improvement is a systematic application of the scientific method, which involves:

1. Observation and recording of facts.
2. Analysis and classification of facts.
3. Development of possible answers and proving them by experiment.

2.5 Work measurement

TIME STANDARDS

The object of work measurement is to obtain a standard time for performing a task. This is expressed in standard minutes (SM) and/or standard hours (SH) and is set on the following basis:

INDUSTRIAL ENGINEERING

1. *For a defined method*   Normally laid down in a document called a 'production method standard' (PMS).
2. *By a qualified operator*   One who is skilled and proficient at the task.
3. *At a specified performance*   The rate of output which the worker performs over the working day. A specified efficiency level.
4. *Inclusive of contingency and rest allowances*   These are allowances for occasional elements of work which are non-repetitive, plus allowances for personal needs and to overcome fatigue.

Time standards are needed:

1. To make predictions of accurate delivery dates.
2. To load machines, processes and manual sections to make the best use of available resources.
3. To make it possible to cost each item of work realistically.
4. To provide facts on which sales potential can be compared against budget.
5. To provide realistic estimates for new work.

TECHNIQUES OF WORK MEASUREMENT

These are:

– Time study
– Activity sampling
– Synthetics
– Predetermined motion time system (PMTS)
– Analytical estimating
– Comparative estimating
– Multiple regression analysis (MRA)
– Computerized planning and estimating

2.6 Properly applied work study leads to higher productivity

It had been thought that as automation took over more work from humans, work study would not be needed. In fact the reverse is true. Work study needs to be applied to automation work in order to make it truly effective.

Work study is also needed in the conception and planning of the project. Techniques such as network or critical path analysis, critical examination and flow charting are all appropriate.

2.7 Summary

Higher productivity provides opportunities for raising the general standard of living as follows:

1. Larger supplies of both consumer and capital goods at lower costs and lower prices.
2. Higher real earnings.
3. Improvement in working conditions.
4. Strengthening the foundation of human wellbeing.

Do not forget the human factor in work study—we are dealing with individuals and we are all different and react in different ways.

3

Historical development

3.1 Introduction

The manufacturing plant we know today has evolved from very humble beginnings made only a few hundred years ago. Until that time most goods or services were produced by workmen at home using their own tools. The worth of the product was measured by the price they charged.

The Industrial Revolution is generally stated to have begun in 1776, which date saw the publication of Adam Smith's *The Wealth of Nations*, part of which dealt with the division of labour; Smith stated that the 'greatest increase of quantity of work is owing to three different circumstances...', namely:

1. Improvement in the dexterity of the workman and increased quality of work so produced.
2. The advantage gained by saving time in passing one sort of work to another is greater than imagined. [Because the subdivision of work leads to more specialization.]
3. How much labour is facilitated and abridged by the application of proper machinery.

3.2 Scientific management

As early as 1832, Charles Babbage, the English mathematician and inventor, voiced concern over 'uneconomic use of men and machines in the production of goods'. His view was that scientific methods should be applied to the production of goods. However, Frederick Taylor of the Midvale Steel Works in the United States is generally regarded as the 'father of scientific management', and although some of his concepts were very unpopular many of them remain with us in industry to this day. His contributions have been listed as:

1. Use of scientific methods of approach in the analysis of management problems.

2. Systematized experimental approach to solving production management problems, as illustrated by his pig-iron handling experiment and his shovelling experiment.
3. Discovery of high-speed tool steel.
4. Monumental work on the art of cutting metals.
5. Development of time study aimed at determining a 'fair day's work'.
6. His differential piece-rate system, a new approach to the problem of designing incentives for work.
7. Development of the concept of staff specialists in industrial organization.

In 1911 Taylor published his *Principles of Scientific Management*. One who worked with him and was influenced by his work was Frank Gilbreth. He refined motion and time study, and together with his wife Lilian Gilbreth developed the application of micro-motion techniques and made an important contribution to the study of fatigue.

In 1925 A. B. Segur developed these principles further with the system of 'Motion Time Analysis'. MTA was a 'predetermined motion time system' (PMTS) of obtaining work standards.

This was followed in 1934 by another early PMTS system, that of 'Work Factor' by J. H. Quick; but the most widespread system in the world, and the most universally recognized system is that of 'Methods-Time Measurement' (MTM), developed in 1946 by Maynard, Stregmerten and Schwab.

The MTM system has a detailed data card of basic motions (reach, move, grasp, position, release, body, leg and foot motions, etc.), each concerned with particular variables. Basic motions are identified, and, with the variables considered, the appropriate times are chosen from the data card. Each element of motion has to be considered in great detail and is slow to apply; while the exactness and minute breakdown of motion can lead to problems of applicator error and inconsistency. Synthesized versions of MTM have been developed to reduce applicator error and the time of analysis, two of them being MTM-2 and MTM-3, which average or group together certain basic motions to produce 'get' and 'put' and so simplify the system and reduce applicator error. These concepts of analysing work are used to an increasing extent by work study and industrial engineers today.

The decades after the Second World War saw the development of the technique of work study. This is primarily concerned with manual work, and to a lesser extent with work performed by machines, but is hardly ever concerned directly with mental work. Work study was really a combination of two main activities, method study and work measurement. In 1959 the British Standards Institution published its British Standard 3138: *Work Study*, which although not recognized throughout the United Kingdom has gained a high reputation overseas where it is often officially recognized; and in

1969 Ralph Barnes published a book, *Motion and Time Study*, from which the work study techniques were refined and developed.

3.3 Social and psychological aspects

The reactions against scientific management have been many, and particularly in the 1960s various attempts to refute these principles were made by those involved in 'behavioural' schools of management. The social conditions of work and different concepts of work organization were investigated, among them job design, job enrichment, job enlargement, job restructuring and group or team working concepts; many of these are important considerations for industrial engineers.

The psychological and sociological behaviour patterns of individuals have been studied for many centuries. Analysis of a more scientific kind was perhaps instigated by Freud, Adler and the *'Gestalt'* psychologists. The Hawthorn experiments in the United States are also seen as scientific attempts to understand sociological aspects of work. More recently there has been considerable research work, eventually evolving into the so-called 'behavioural' or 'human relations' school of management.

In 1954 the American psychologist Maslow differentiated between various human needs and aspirations. His theory of motivation comprises a hierarchy of five levels of needs, as listed below, the individual beginning by satisfying the first basic need and then, as this is satisfied, going on to satisfy a higher level of need, thus:

1. *Physiological needs* These are the essentials for survival, such as food, water, air and sleep.
2. *Safety/security needs* The desire for protection and security from danger.
3. *Social needs* The needs for belonging and for love and affection.
4. *Esteem needs* These are twofold:

    (a) Self-esteem, which is the need for self-respect.
    (b) Esteem from others is the need for others to give respect.

5. *Self-actuation* The need for a person to reach his full potential, i.e., self-fulfilment.

These constitute the intrinsic motivation of an individual, or motivation stemming from within. Also extremely important is extrinsic motivation, that which stems from factors outside, for example quality of leadership, organization of work environment, welfare amenities and pay.

McGregor was influenced by Maslow's work, and argued that what managers assumed about a person's behaviour at work was based in turn on

Table 3.1   Summary of McGregor's Theory X and Theory Y

Assumptions about man

Theory X	Theory Y
Man dislikes work and will avoid it if he can.	Work is necessary to man's psychological growth. Man wants to be interested in his work and, under the right conditions, he can enjoy it.
Man must be forced or bribed to put out the right effort.	Man will direct himself towards an accepted target.
Man would rather be directed than accept responsibility, which he avoids.	Man will seek and accept responsibility under the right conditions. The discipline a man imposes on himself is more effective, and can be more severe, than any imposed on him.
Man is motivated mainly by money. Man is motivated by anxiety about his security.	Under the right conditions man is motivated by the desire to realize his own potential.
Most men have little creativity—except when it comes to getting round management rules!	Creativity and ingenuity are widely distributed and grossly underused.

some background assumptions. These he divided in terms of 'Theory X' and 'Theory Y', summarized in Table 3.1.

Another person who sought to improve our understanding of the nature of our reaction at work was Hertzberg. He developed his 'Two Factors Theory' ('Motivation and Hygiene Theory') in which he determined those things which gave job satisfaction and those which were dissatisfiers. The two were not opposites: the things which caused dissatisfaction are things which tended to switch people off. The satisfiers were described in terms of job enrichment (sometimes referred to as job redesign).

These theories are now seen to be a simplistic approach, but are often useful in understanding certain features of human endeavour. Man is, on the whole, an extremely complex being, with a wide variety of goals, attitudes, backgrounds, education, beliefs, cultures and so on. Certainly, many have questioned the idea that increased job satisfaction increases productivity. Vroom, with his 'Expectancy Theory' (1964) of probabilities of desired outcome, is one such. He undertook research studies which showed no statistical relationship between job satisfaction and performance. Porter and Lawler reversed the cause-and-effect relationship, arguing that good

INDUSTRIAL ENGINEERING

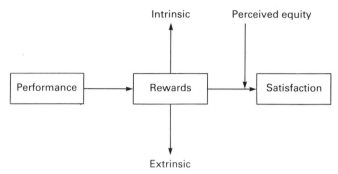

Figure 3.1 Porter and Lawler's model of the relationship between performance and satisfaction

performance which elicits rewards causes job satisfaction. Their model is shown in Fig. 3.1.

Other studies of work behaviour were undertaken by the Tavistock Institute of Human Relations, who adopted what is known as a 'socio-technical' approach, finding that both social and technical factors affect job performance, while both are independent yet they do affect one another.

A large amount of research and publication has been undertaken on the human aspect of work. Much of this provides insights into certain aspects of behaviour, but it is always worth testing the underlying assumptions behind any theory or philosophy to elicit what it is they are really about. The next chapter summarizes some of these concepts.

4

What is management?

4.1 Introduction

We often have a better appreciation of what we are talking about if we can define our terms. Definitions need to be precise, but unfortunately this does not mean that everyone will have the exact same understanding, since at best they only give an approximation of the original thought.

A definition of management is 'a creative activity which develops an organization and its people to achieve specific goals'. This needs 'unpacking' (What is meant by creative activity? How do we develop organizations and people? What is meant by specific goals? etc.), but it is sufficient to provide an approximation.

'Creative' here implies innovative design, bringing forth new thoughts and acquiring relevant knowledge.

The goals of an organization may be any of the following examples as overall objectives, but each one can be broken into many sub-objectives:

- The provision of products or services to the community.
- The economic and social wellbeing of a nation.
- The search for and dissemination of knowledge.
- The health care of people.
- The worship of God.

Each of the organizations performing these tasks will be very different from the others, as they could be, typically, a manufacturing concern, a government, a university, a hospital complex, and a church. Management and organization are necessary in each case. The more that goals and objectives are specific, the better.

Management styles within an organization differ according to the personality and temperament of each individual manager and are influenced greatly by culture and beliefs. Management is often thought of as having evolved from 'scientific management', the oldest concept, through 'behavioural management', then 'systems management', and so on. Perhaps the historical perceptions of management have come about in this way; but it is our belief that each of these different aspects of approach to management is relative and needs to be integrated with the others.

Without scientific principles, management is no more than acting on beliefs and feelings; these may be relevant, but in today's high-technology society they are hardly sufficient. Beliefs can be very different between individuals, and feelings are apt to change from day to day.

4.2 Scientific management

Scientific management implies analysis, measurement and methodology. It is essential in modern organizations in gaining insight and knowledge of a situation. The knowledge may not be total or the insight complete, but without information of this kind decisions tend to be made with inadequate data and often the worst kind of decision is made. Some managers have been accused of adhering to the motto 'My mind is made up; don't confuse me with facts!'

Scientific knowledge is an important facet of knowledge, which provides greater insight. Knowledge has many facets, according to where we start and how we look at events and things. Physiologists tell us that sight can be conceived of as sensory perception through the cornea, through the cones and rods of the retina to signals received in the brain which are then interpreted by association. On the other hand, we are told by the same physiologists that we have to have some mental image to start with, and therefore that two-way interaction between brain and eye messages is needed. Further, we can see only what we expect to see; the eyes are merely interpreters of what we already sense.

Scientific knowledge is limited to that which is measurable or quantifiable. It is only part of total knowledge and is quite limited, but it is most essential. Measured knowledge is not creative until that knowledge is 'synthesized', or built into a coherent whole; this is one aspect of management. Science and the use of science (technology) has given us many advantages during this century—and some disadvantages. Thus scientific knowledge can be used for good and evil, and discernment and judgement based on belief are necessary.

Science is based on belief—that we are living in an ordered and rational universe, and that there are natural laws to be discovered. The authors' belief is that companies would operate much more effectively if they were much more specific and scientific in their approach and outlook as regards business life: measurement does give greater depth to understanding.

4.3 Behavioural management

This is based on the fact that an organizational activity is a human enterprise, and that work is directed to human goals—certainly both manufacturing and service organizations serve human needs:

1. Humans are customers, therefore the products, produce or community service offered by an organization would not be necessary without human beings.
2. Work provides employment. Most of us have to work in order to live; some less fortunate have difficulty in obtaining work, occasionally because they have not wanted to adapt to new environments and skills, and more often because of trade recession, when whole classes of work have been made redundant.

Work has two advantages:

1. It provides a means of satisfaction to most people, who take pride in what they achieve. The craftsman working with his hands has many advantages over others in this respect, as has the person who serves others, such as in health care. Other people's work can also be intrinsically interesting—this is perhaps the area of the support service professional who is engrossed in others' work where study, observation and practice are all involved with a desire to continually improve.
2. It provides the means of living, giving us the medium of exchange for our daily bread. Too many people today have over-emphasized this aspect of work, where the material rewards far outweigh the input. There is nothing wrong in having possessions, provided they do not possess us. Nothing is worse than the professional who is 'in it for the money'—especially those in the stock market, law, banking and finance whose aim is often to exploit those he should serve: 'Man shall not live by bread alone.'

Behavioural management bases its knowledge and conclusions on the findings of psychology, physiology and sociology, which in turn derive from theories, empirical research and interpretations. Such findings are useful in that the patterns of behaviour of social groupings can be known, and used 'creatively' for furthering the goals of the enterprise; they do not, however, show how individuals or groups will always behave in the future: each individual is different and needs to be treated as such, and is to some extent unpredictable. Management has primarily to manage individuals; if it is 'creative' it will usually be interesting, stimulating and promote dedication to goals.

Perhaps one crucial question to ask is whether people are born with individual characteristics or skills. If so, what do we inherit genetically or through the physiology of our parents? Also, are people trained or conditioned by their early life? And which skills and abilities are provided after birth, which through continuing education and training later in life? As far as managers are concerned, the old question continues to be asked: 'Are managers born or made?' Our view is that we have been influenced greatly by all the outside factors, but that perhaps the basic behaviour patterns are there from a very early age, while crafts, skills and educational knowledge can be acquired throughout one's life, even to a relatively old age; some are better able to continually acquire such knowledge and skills than others.

What management has to do is to be creative in recognizing and selecting persons for the most appropriate tasks or job—in terms of creative ability, skills, development, potential and human relationships. If a manager does not recognize appropriate abilities and develop these, through training and experience, to the highest level possible, then he is not making the best use of his resources, and is failing as an effective manager. The ability to lead people carries with it the ability to relate, confer, develop, motivate and be trusted, together with impeccable truthfulness and honesty.

4.4 Systems management

Many have advocated what is called the 'general systems theory' applied to management. A system has been defined as an 'organized or complex whole or an assemblage or combination of parts forming a complex or unitary whole'. This certainly needs 'unpacking' for a more complete understanding. An idea of what a system is can come from having studied scientific subjects, and most people have some understanding of the galactic systems of the universe, of geographical systems and of molecular systems. Anyone who has worked with computers for a time will understand what computer systems are meant to be. We also hear of, and experience daily, transportation systems, telephone systems and economic systems. The word 'system', in fact, is being used more and more to refer to more and more facets of our society.

Many systems are 'closed' systems, some of which are viewed generally as mechanistic—clockwork systems, ignition systems, brake systems—while others are of a more complex type, often referred to as 'feedback' systems and including notably self-regulating types such as thermostatic controls, industrial process controls, and the 'environmental' systems for temperature, humidity and air particle control.

The other kind of system is the 'open' system. These are often defined as 'self-maintaining structures'; they start with the cell as the most elementary form and progress through to genetic–societal as typified by the plants—vegetables, flowers or trees—to animals—with their increased mobility, teleological behaviour and self-awareness—and then to humans—with their ability to use complex language, to communicate, think and remember, and to visualize 'symbolically'. Social systems or systems of human organization come next; these include all forms of cultural system. The supernatural and moral laws are often classified as transcendental systems.

Note that while the concept of a system may seem to imply '*laissez-faire*', in the sense that somehow it can be left alone to 'take care of itself', this can be a very dangerous assumption. Any organization or system needs good leadership: without leadership, it will deteriorate; with poor leadership it will also deteriorate but over a longer period.

4.5 Developing management skills

Although a complete book could be written on the subject, it must suffice here to state that the main factors contributing to the development of management skills are as follows:

– Communications
– Leadership skills
– Team working
– Decision making
– Delegation
– Motivating
– Improving personal relationships
– Counselling/coaching
– Time management

LEADERSHIP

This is very difficult to define. Simply, it can be stated to be 'those attributes of personality and character which get the best out of people'. However, any attempt to define the detailed attributes, such as the many already made, could result in a list of many thousand variations. Behavioural scientists have tried to define leadership by techniques such as 'temperament analysis', but other characteristics are involved too, such as strength of character, moral attributes (honesty, truthfulness and belief), the ability to get on with people, and integrity (the quality which makes people trust you).

Three basic kinds of authority

1. Authority of position   Job title, rank, appointment.
2. Authority or personality Natural qualities of influence.
3. Authority of knowledge   Technical or professional.

For instance, leadership is required to conduct the London Symphony Orchestra; not all of us would be qualified for that—even if we were good leaders. The field of activity is relevant to the particular leadership needed for a specific position.

INDUSTRIAL ENGINEERING

Figure 4.1 Action-centred leadership

Action-centred leadership

This is a concept introduced by John Adair in his books *Action-centred Leadership* and *Leadership*. Its focus is on the task goal, and the means of achieving it are in the three areas given in the circles in Fig. 4.1. It also looks at:

1. The leader needed — Personality and character for the situation.
2. The situation — What are the practical considerations involved?
3. The group — What are the expectations, needs and wishes of the people involved?

The goals and primary tasks for the leader need to be expressed in the terms given in Fig. 4.2.

Figure 4.2 Goals and primary tasks for the leader in action-centred leadership (Source: Adair: *Action-centred Leadership*, Gower, 1979)

5

The basis for success—the human factor

5.1 The worker

An organization's most important resource is its people. This resource can make or mar any venture. It is a resource which is readily available, which can appreciate in value, can be rapidly developed, and is the only one which can make a significant impact on an organization's success.

All people, whether they are managers, administrators, technicians or operators, are workers. They each have potential which can be readily identified. A good worker is one who:

1. Has the natural abilities and specialized training to do a particular job (or jobs) well.
2. Is highly motivated with regard to quantity, quality and cost.
3. Is a positive contributor to the organization, at whatever level his abilities are best suited.
4. Is stable and mature, but also has imagination and flair.
5. Is able to contribute as a member of a team, with the attitude of wanting to provide a service.

THE ABILITIES AND TRAINING TO DO THE JOB WELL

'No one is good at everything, but everyone is good at something.' Natural abilities and talents should be used in an organization in the best possible way. In fact, most people have a variety of abilities which can be expressed in the work situation. Most people would admit that they are not being used to the maximum potential.

A good worker, to perform his job well:

1. Will be aware of his own shortcomings and will take steps to overcome these.
2. Will use vision and imagination to improve what already exists.

3. Works smart: uses his brain to work efficiently.
4. Performs to high standards of quality and reliability in his work.
5. Manages his time well.

HIGH MOTIVATION WITH REGARD TO QUANTITY, QUALITY AND COST

Motivation is 'get up and go', and the best aims to which it is directed are the achievement of success in the task, personal enhancement, and the development of the group of people closely associated with the task.

A highly motivated worker:

1. Sees things to be done and takes action to do them.
2. Responds to a challenge and likes solving problems.
3. Takes an interest in the task and the people closely involved.
4. Gains satisfaction from achieving.
5. Is interested in setting goals for attainment and makes sure that they get there.

A POSITIVE CONTRIBUTOR TO THE ORGANIZATION

At whatever level, such a person will seek to promote the organization, themselves and their colleagues, so that they can make an even more important contribution in the future at a higher level.

A positive contributor:

1. Starts from the bottom and works up, and is prepared to offer a good service to others at all times.
2. Has high standards for his work.
3. Organizes his workplace well and sees that it is kept tidy and clean.
4. Is reliable, accurate and consistent.
5. Is able to work under authority.
6. Can change to different demands of the work situation.

STABILITY, MATURITY, WITH IMAGINATION AND FLAIR

A person possessing these:

1. Has moral integrity and truthfulness.
2. Has a strong sense of responsibility.
3. Will speak up if something is wrong and will not give way under pressure to relax standards.
4. Will speak honestly face to face, but with heartfelt concern for other people.

5. Responds but channels ups and downs in his emotions to positive ends.
6. Learns from experience.
7. Has a strong urge to develop spiritually, mentally, emotionally and personally.

ABILITY TO CONTRIBUTE AS A MEMBER OF A TEAM

A team is a number of people associated together in work or other activity. Most people want to participate and to be part of a group, whether this means supporting Manchester United or belonging to the local darts team. We mostly all seek the companionship of our fellows. At work the same attitude should be allowed to prevail. Most people would like to feel part of the organization, to know about what is going on, to participate in its growth and rewards.

A good worker:

1. Will get on well with others.
2. Will be able to communicate, both by the spoken and written word and by expressions.
3. Will have the ability to listen.
4. Works well in team situations.

5.2 The human factor

Needless to say, although a person may be trained in all the techniques of industrial engineering, the human factor plays an important and significant part in its success, and indeed previous chapters were written so that this fact is constantly remembered.

The industrial engineer must seek as a main objective to promote and continue good relationships with those he has to work with. If the people he is working with as well as the trade union representative are not taken into account, disastrous results will usually follow. It must be stressed of course that these human relations goals are not sought at any cost, for instance at the sacrifice of integrity and truth: this would result in the loss of goodwill and respect from all working colleagues.

It is also important to emphasize that industrial engineers are *not* line managers. They cannot give direct orders to other people: that is the supervisor's job. It is important to discuss with supervisors the various aspects of industrial engineering which will impinge on their own working environment. Supervisors should be 'built up' rather than 'pulled down', because each has the authority of his position and can be a very powerful force for good provided he knows and understands what goals are to be accomplished.

The personal attributes of the industrial engineer are most important. In addition to those of a good worker given above, he should possess the following:

1. *Education*   The IE should preferably have received formal education beyond the secondary school, i.e., a university degree or polytechnic or college diploma, followed by professional training in industrial engineering or in work study and management services. There may of course be exceptions, but the level of knowledge and training required for an industrial engineer is higher than would normally be required for a work study engineer.
2. *Practical experience*   It is desirable that the IE should have had experience in more than one type of industry and should have received some formal on-the-job training in industrial engineering practice with a more experienced professional, prior to practising alone. Graded experience over a two-year period is necessary before complex projects can be undertaken.
3. *Personal qualities*   These are essential: they include things such as a good personality, sincerity and honesty, enthusiasm, interest and empathy with people, tact and understanding.
4. *Good appearance and attitudes*   Industrial engineers should be efficient people themselves. The methodical way they work, the good organization of their own workplace, and their manner and dress should reflect the role and importance of the competent professional. Formal executive attire, well kept, is most necessary.
5. *Confidence*   The industrial engineer should possess the necessary confidence to perform his task well. Factors listed in points 1 to 4 above will tend to help with this particular attribute. He must be able to win the support and confidence of others.

Part 2
Methods engineering

6

Methods engineering

6.1 Introduction

Methods engineering is concerned with facility design and work design methods. 'Engineering' is used in the sense of 'engineering something', rather than in the mechanical engineering sense, so it is applicable to any form of work, anywhere.

Facility design is concerned with how smoothly work and materials can flow; it is also pertinent to the human interaction within this environment (both ergonomic and psycho-sociological aspects) and how well the organization can adapt to changes in methods of working, or to changes in the product or in the service being performed.

Work design is more concerned with how individual tasks or groups of tasks are done, and thus with the use of properly designed work stations, with the way materials are transported and set out for the task, and with whether hand-assisted tools or automatic methods are needed. Although work design is traditionally linked more to the human-task environment, work is increasingly being performed by power-assisted manipulation and feed mechanisms, which rely less on direct operator involvement.

Before any design work is carried out, it is necessary to establish the goals and policies for the operational task, which may be manufacturing but can be any kind of operations environment, also to assess the flows and tasks to be done in a methodical and systematic manner and to analyse in detail what is needed. The analysis could be of *existing* methods and flows, or of *potential* methods and flows. It involves building a 'model' of what the real-life situation looks like or will become.

6.2 Modelling

Modelling is used with increasing frequency in operations. A 'model' is a representation which portrays the key features of the situation being investigated. It breaks down the whole into an array of sub-features, which are used in order to gain an understanding of the situation. These sub-features usually consist of the following kinds of model:

INDUSTRIAL ENGINEERING

1. *Chart*   Shows movements, events and flows over a period of time. It is obviously a static picture of moving events.
2. *Mathematical*   Here the flows and measurements can be expressed by mathematical equations or by statistical analysis to show causes and their probable effects.
3. *Financial*   This is really a model where quantities have been replaced by values or costs.
4. *Simulation*   In this type of model the events are represented by a combination of the above, with the results often portrayed in picture form, but shortening the time-span. For example a whole year's events may be portrayed in five minutes.

Points 1 to 4 can be performed manually. However, the computer is increasingly being used to represent all the types of model mentioned. The more complex the study or the more complex the model, the more likely it is that computers will be necessary. For instance, there are many simulation packages available for computers, including the following types:

– KBS (Knowledge Based Simulation) programs developed by Carnegie-Mellon University in the United States.
– FADES (Facility Design Expert System) developed by Purdue University in the United States.

There are also many companies which offer simulation software and programs, e.g., McDonnell-Douglas, General Electric, Panasonic, Arthur D. Little, Boeing and IBM in the United States, and PE Consultants, Pactel, Istel (where the Metro car line was simulated prior to production) and Ingersoll Engineers in the United Kingdom.

There are many financial microcomputer packages available for modelling purposes; these are often called 'spreadsheets' because they are arrays of columns and lines of figures and values with their relationships stipulated by simple statements through the program itself. There are also more powerful computer systems for financial modelling.

Industrial engineers should use their skills and ingenuity in order to develop the most appropriate models for their task, whether computerized or manual. Simple models may often be the best way of visualizing and collecting data to begin with; it is usually only to obtain more accuracy or a better representation that complex models are used.

PROCESS CHARTING

The purpose of a chart is to record facts in a systematic manner; a sequence of events is portrayed diagrammatically by means of a set of process chart

METHODS ENGINEERING

symbols; these aid the visualization of a process as a means of examining and improving it. The symbols used are as follows:

Charting symbols (OTIDS)

Operation A main step in a process, method or procedure.
○

Transport Movement of works, materials or equipment from place to place.
⇨

Inspection An inspection for quality and/or check for quantity.
□

Delay A delay in the sequence of events, for example work waiting between consecutive operations or any object laid aside temporarily without record until required.
D

Storage A controlled storage in which a material is received into or issued from stores under some form of authorization, or an item is retained for reference purposes.
▽

Types of charts

There are five main kinds:

1. Recording charts.
2. String diagrams.
3. Travel charts.
4. Proximity charts.
5. Logistics charts.

Recording charts

These comprise the following types:

1. *Process chart* Chart in which a sequence of events is portrayed diagrammatically by means of a set of 'process chart symbols'; as already mentioned, this assists the visualization of a process as a means to examining and improving it.

INDUSTRIAL ENGINEERING

2. *Outline process chart*   A process chart giving an overall picture by recording in sequence only 'operations' and 'inspections'.
3. *Flow process chart*   A 'process chart' setting out the sequence of the flow of a product or a procedure by recording all events under review using the appropriate process chart symbols. Flow process charts may be of the following types:

   (a) *Man type*   Records what the worker does.
   (b) *Material type*   Records what happens to material.
   (c) *Equipment type*   Records how the equipment is used.

4. *Two-handed process chart*   A process chart in which the activities of a worker's hands (or limbs) are recorded in relationship to one another.

From / To — Distance (metres)

To \ From	Prod. planning	Inspection	Main stores	Tool stores	Purchasing	Work study	Costing
Prod. planning		2.8	3.8	2.8	4.2	10.0	4.0
Inspection	2.8		3.6	4.5	5.6	4.7	4.8
Main stores	3.8	3.5		25.0	20.3	2.5	4.8
Tool stores	2.8	4.5	25.0		5.7	3.5	4.8
Purchasing	4.2	5.6	20.3	5.7		6.6	12.5
Work study	19.9	2.2	3.5	2.5	6.6		8.7
Costing	4.0	4.8	4.8	4.8	12.5	2.7	

Figure 6.1   Travel chart, distance type

METHODS ENGINEERING

5. *Multiple-activity chart* A chart on which the activities of more than one subject (worker, equipment or material) are each recorded on a common time-scale to show their interrelationship.
6. *Simultaneous-motion cycle chart ('Simo' chart)* A chart usually based upon 'film analysis' to record on a common time-scale the 'therbligs' or groups of 'therbligs' performed by different parts of the body of one or more workers. (A 'therblig' is a symbol devised to represent very minute elements of work.)

String diagrams

In these a thread is used to trace and measure the routes, which are laid out on a scale drawing of the building or facilities.

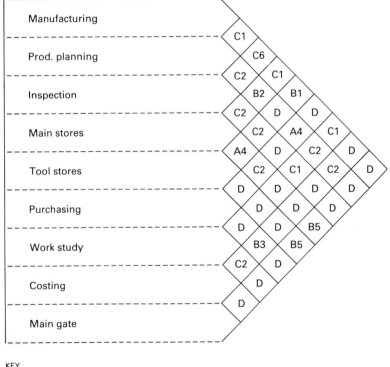

KEY
Proximity Must be adjacent	A	*Reasons* Much personal contact	1
Closeness essential	B	Much transfer of paperwork	2
Closeness important	C	Use of common equipment	3
Not essential to be close	D	Continuous liaison	4
Must not be adjacent	X	Free access to roadway	5
Must not be even close	Z	Must be clear of fumes	6

Figure 6.2 Proximity chart

Part No.	Description	Unit of measure	Quantity	Frequency and distance moved	Container and size	Method of transport	Method of storage

Figure 6.3 Logistics chart

METHODS ENGINEERING

Travel charts

Travel charts are useful for summarizing information on flow lines, such as frequencies of journeys, or distances between locations. The types of chart are:

1. Distance type (see Fig. 6.1).
2. Frequency type.
3. Multifactor type.
4. Proximity type.

Proximity charts

See Fig. 6.2 for a typical specimen. As an example of its use, closeness is essential because the same planning board is used: B3.

Logistics charts

When process charts or travel charts are used the logistics chart is extremely useful (Fig. 6.3).

Whenever materials are transported, containerized or stored, those three features impact on one another. For instance, the quantity in a batch will often determine how they are grouped (containerized), and this in turn determines how they are transported and stocked. A small quantity of a light article (e.g., pins) will normally be placed in a small box (carton) and transported manually (by hand); they will often be stored in the small box (carton) in an open storage arrangement (shelf). A large quantity of the same article (pins) will still be placed in a series of small boxes (cartons) which will be placed on a pallet, and they will be transported by vehicle (forklift truck) on structural supports (tubular structure for supporting pallets and larger containers). The batch quantity is an important one and it is necessary to distinguish between the following:

1. *Process batch* The quantity the process will normally provide in a given run or for a given order.
2. *Movement batch* The quantity desirable for movement. This will often be less than the batch size and may even be one on an assembly track.
3. *Storage batch* This will often be related to container size and method of storage.

7

Principles of methods engineering

7.1 Introduction

Ryuji Fukuda in his book *Managerial Engineering* (Productivity Press, 1983) has provided some useful insights employed for methods engineering in Japan. These are techniques for improving quality *and* productivity at the workplace. The two should always be considered together with *costs*.

Success in doing things well is defined by the following cycle of activities:

1. Develop a reliable method.
2. Create a favourable environment.
3. Keep every worker practised in the method.

Methods which can reduce defects by 50 per cent and yet yield returns of between 30 and 50 per cent additional productivity should be carefully examined. This is achieved mainly by groups of people working together. The favourable environment is one in which people can work as a team (as in a quality circle), by using modern techniques of management to examine and develop new *reliable methods*, and by keeping people well used to the work for which they have been trained. (If we keep repeating something we get better at it, if we don't practise we lose the skill.)

Examples of reliable methods improvements are the following:

1. Use of cause-and-effect diagrams (see Fig. 7.1) to improve quality and remove sources of unfavourable causes of poorer performance at later stages.
2. Identification of the causes of accumulation of work in progress and the control of flows through stockless production, often called 'just in time' techniques (JIT).
3. Reduction of setup times to well below their original cost (Tables 7.1 and 7.2 show how this was done).

PRINCIPLES OF METHODS ENGINEERING

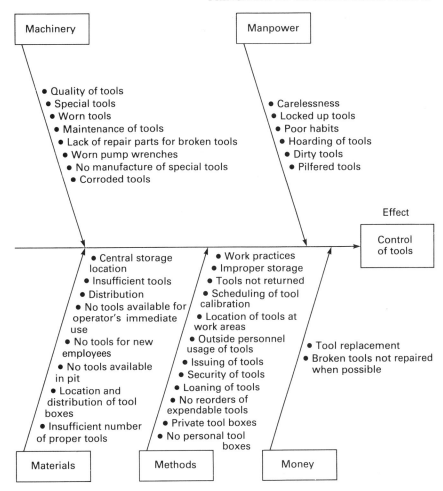

Figure 7.1 Cause and effect diagram

7.2 Methods applied to facility layout and design

Good facility layout is essential for operations of all types. It comprises the prime activity from which all other method engineering principles derive. The layout of materials and process flow affects how workplace layouts are made. Facility layout is the key to productivity and quality at all sub-stages. For example:

1. How hospitals are laid out with regard to major functions will determine how efficient specialist functions are. For instance, if wards are a long

Table 7.1 Steps in reduction of setup times

Step	Description
1. Choosing a setup operation	(a) Set the objective clearly: e.g., shorten the setup time by 80 per cent for: – small-lot production; – reducing operators; – expanding production capacity. (b) Choose an operation which greatly needs improvement: e.g., operations which: – take a lot of time; – are very tiring; – involve difficult adjustments; – are very frequent. (c) Choose a specific machine rather than a group of machines.
2. Observation and measurement	(a) Measure average mean time in typical cases (a small number of cases will suffice), e.g.: – cases of major products; – cases of products with frequent production. (b) Apply time study for short-cycle operation. (c) Apply operation analysis at 1-minute intervals: – for long-cycle operation, making description at level of elemental operation. (d) Measure operations with higher frequencies at a more detailed level.
3. Summing up	(a) Prepare a summary form and fill it out. (b) Post it on a special noticeboard where it will attract suggestions from many people
4. Analysis	(a) Examine the purpose of each elementary operation and possible results if it is abolished. Examine from as many viewpoints as possible. (b) Use cards: they are effective in attracting and presenting many people's ideas without requiring meetings.

Table 7.1 (continued)

Step	Description
	(c) While examining purposes, try to improve operations that: – require a tiring posture; – require physical strength; – are unpleasant to perform; – interrupt the rhythm of work; – require attention.

5. Idea generation	(a) Hold meetings to gather ideas. Generate as many ideas as possible which can be realized quickly and cheaply even if their effects are limited.
	(b) Utilize hints for reducing setup time:
	  (i) Draw a clear line between internal setup (done while a machine is stopped) and external setup (done while a machine is working).
	  (ii) Change internal setup to external setup.
	  (iii) Apply one-touch fixing and removing.
	  (iv) Eliminate adjustments.
	Develop alternative ideas with 80 per cent effectiveness and 10 per cent cost.

6. Execution	(a) Try the new ideas, easiest ones first. (b) Have manager and engineers examine effects of new ideas on product quality and safety.
7. Maintenance	(a) Measure the result of the improvement using day-to-day management. New ideas may be obtained in this process. (b) Standardize the new operation after it is stabilized, to maintain the achieved result.
8. Repetition	At appropriate intervals, repeat the above steps.

Table 7.2 Results of setup time reduction technique given in Table 7.1

	December 1977	March 1978	August 1978	December 1978
Number of projects completed	121	186	278	308
Average rate of reduction in setup time	40.0%	38.5%	41.5%	40.5%

distance from service departments and operating theatres then staff will spend much of their time travelling rather than doing effective work.
2. The layout of a supermarket as regards the shelves, counters, storage and checkout must be effective for display and selection of goods and flow of traffic, while also taking into account the services needed, i.e., lighting, frozen storage, power supply, heating, etc.
3. For manufacturing units, many new methods are being introduced. Flexible manufacturing systems (FMS), computer integrated manufacturing (CIM) and computer aided design and manufacture (CAD/CAM) will need to be integrated with or changed to from the traditional types of layout such as process layout, product layout, fixed-position layout or group technology cell.

Layout begins with the total strategy for operations or manufacturing. Unless the latter has been thoroughly conceived, it will not reflect the best possible use of resources. The actual location itself may have to be considered initially. Where should major facilities be situated? What is the economic size of these facilities? Should these be specialized units or should they each do the same thing, with duplication of all the major functions? Government grants and local authority backing should probably be the last things considered rather than among the first.

PRODUCT LAYOUT

Typical of layout by product is the assembly line, which is normally employed when goods or services are to be rendered in high volumes. The organization of work is dictated by the sequence of processes or operations which have to be carried out. Examples of this are: automobile assembly, car washes, food processing.

Assembly lines operate typically at a 'forced' work-pace; this is usually dictated by the speed of conveyors or transfer equipment. 'Paced' work is where the work is moving continuously or at frequent fixed intervals;

'unpaced' work is where the operator's time is balanced but work is transferred by belt only when the work is placed upon or linked to it.

Assembly-line balancing is necessary to divide the total work content up equally between the number of operations to be performed. If all workers and machines are 100 per cent occupied, then the line is said to be perfectly balanced. There always has to be some latitude in the way in which an assembly line is balanced because some workers will be better than others. Other variables will also come into the picture; for instance, it is not possible to divide the operation time exactly equally among all the work stations.

The quality of the solution can be calculated by the following formula:

$$\text{Balance delay} = \frac{nc - T}{nc}$$

where n is the number of work stations
c is the cycle time (c cannot be less than the slowest cycle)
T is total work content of product

Computerized systems can be used to balance assembly lines. For instance, computerized MOST provides a line balancing option in its programs.

PROCESS LAYOUT

In this type of layout all the operations of a similar nature are grouped together. These are typically used in jobbing or batched processes. The organization normally offers a wide range of products or services in the market-place, and quantities are not sufficient to justify continuous operation.

The arrangement of processes and the flow of work for different jobs may be against one another. Usually, operators would have at least one job awaiting their attention to avoid under-utilization of equipment and personnel. This waiting of jobs represents a queue, which can be assessed mathematically (queuing theory in operational research) or by simulation techniques (flows into and out of facilities).

Batch size also represents a flow restriction, as well as added value of work in progress and stocks; congestion and long throughputs are also prevalent in this type of layout, unless flows of materials are looked at carefully.

What is really important in process layout is to recognize where bottlenecks occur, usually through key machines or operators that represent shortage of capacity. Work should be organized and planned at the rate of output at the bottleneck, even if facilities before or after are kept waiting.

An alternative to process layout is to undertake a Pareto analysis of the total workload and to plan the layout on cellular principles for the small number of jobs which comprise the larger volume or value.

Computerized systems for analysing process layouts are:

- CRAFT (Computerized Relative Allocation of Facilities Technique), Triangle Universities Computational Centre, North Carolina, USA.
- CORELAP (Computerized Relationship Layout Planning), Triangle Universities Computational Centre, North Carolina, USA.

Further lists of programs can be obtained from the excellent book by R. Wild, *Production and Operations Management* (Holt, Rinehart and Winston, 2nd edn, 1983, pp. 106–7).

FIXED-POSITION LAYOUT

Here the product is in a fixed position and the facilities move round it owing to its bulk or its weight, or because it is not possible to move it once it is built. Typical of fixed layouts is shipbuilding (there are some assembly-line type ship layouts in Japan) and large aircraft assembly; civil engineering projects, building and possibly mining would also come under this category.

7.3 Procedure for layout

NORMAL SEQUENCE

The sequence in planning a layout of facilities is an extremely complex one. It would normally be as follows:

1. Assess market demand and decide on policy for operations.
2. Determine work methods, standards and capacity.
3. Lay out the building, or alterations to the existing one, through

    (a) An architect for new buildings.
    (b) Local authority planning permission.
    (c) Agents, surveyors and legal agreements.

4. Decide on resource requirements according to specification, size and shape of equipment.
5. Handling and movement: analyse, through process charts and logistics charts, the handling, movement and storage facilities required.
6. Design layout

    (a) make general arrangement of buildings.
    (b) Make detailed arrangement of equipment.
    (c) Make projected layout from analysis.

7. Design services: these need to be considered at stage 6 and related to it: power supply, lighting, heating, compressed air, oil supplies, etc.

PRINCIPLES OF METHODS ENGINEERING

METHODS FOR PLANT LAYOUTS

When making plant layouts, there are three different methods which may be applicable, as follows:

1. *Computer-aided design and drafting (CADD)* There are many CADD packages for two-dimensional work which can be used for plant layout work. They proceed as follows:
   (a) Generate the general layout of the building, including columns, windows, office and storage areas, etc. Store in disk file.
   (b) Generate shapes for each type of facility and store, each on disk file.
   (c) Using results of flow analysis, call up shapes from file and place in position in the building using coordinates.
   (d) Print out layout and file completed layout in disk storage.
2. *Traditional drawing-board technique* The procedure is similar to method 1:
   (a) Draw up general layout on tracing paper.
   (b) Draw up shapes of individual facilities on small pieces of tracing paper.
   (c) Using results of flow analysis, place shapes on small pieces of tracing paper on general layout, stick down with double-sided Sellotape.
   (d) Print out layout and file.

   For any changes to the layout, remove layout master from file, make up any new shapes required, unstick changed shapes, add new ones, etc.
3. *Using models of facilities* The steps are:
   (a) Purchase proprietary board; make three-dimensional model of walls, columns, etc., to a height representing 1 metre height (extra if needed).
   (b) Purchase proprietary shapes for machines and equipment.
   (c) Using result of flow analysis, lay out facilities. This method can be used only for demonstration. Methods 1 and 2 are better for filing and copying.

For layout purposes, a modern technique called 'production flow analysis' (PFA) is used to analyse the flow patterns so that groups of work can be linked together.

EXAMINATION

Questioning technique

This term refers to the means by which the critical examinations are conducted, each activity being subjected in turn to a systematic and progressive series of questions. An example of a critical examination sheet on which the questions are printed in sequence is shown in Fig. 7.2.

Element description .... Date ....
.... Reference ....

The present facts		Alternatives	Selected alternative for development
Purpose *What* is achieved?	*Is it necessary?* ☐ YES *If yes — why?* ☐ NO	*What* else could be done?	*What?*
Place *Where* is it done?	*Why* there?	*Where* else could it be done?	*Where?*
Sequence *When* is it done?	*Why* then?	*When* else could it be done?	*When?*
Person *Who* does it?	*Why* that person?	*Who* else could do it?	*Who?*
Means *How* is it done? (Method)	*Why* that way?	*How* else could it be done?	*How?*

Figure 7.2 Critical examination sheet

Creative thinking

This is a philosophy based on psychological principles which examines the way in which creative thinking takes place, and uses these aspects to bring about improved methods, for instance brainstorming techniques.

PRINCIPLES OF MOTION ECONOMY

These are used for the design of the workplace where manual operations are performed:

1. *Minimum movements*   Movements which, while natural, are the minimum needed for the job.
2. *Simultaneous movements*   Movements in which different limbs are working at the same time.
3. *Symmetrical movements*   Movements so arranged that they can be performed on the right and left sides of the body about an imaginary line through the centre of the body.
4. *Natural movements*   Movements which make the best use of the shape and arrangement of the parts of the body involved.
5. *Rhythmical movements*   A sequence of movements which induces a rhythm when repeated.
6. *Habitual movements*   Movements designed, through precise repetition, to become a habit.
7. *Continuous movements*   Movements which are smooth and curved and which avoid sharp changes of direction.

In modern practice, the techniques of PMTS are increasingly used to reduce movements (see p. 62).

7.4 Fallacies being taught

There are many aspects of production theory which are just not true. They are usually perpetrated by those who have no real practical experience of working on the shop floor, and are as follows:

1. *'The assembly line is dead.'*   A properly designed assembly line, with work aids and power-assisted tools, is at the present time a most effective way for assembling products. It will be superseded eventually by lines with robot assembly machines, something that is already happening with the assembly of electronic components to PCBs. Many companies which tried other assembly methods such as group work have now reverted to the assembly line methods, e.g., Volvo in Sweden.

2. *'Cycle time cannot be less than 1 minute.'* It may be necessary for the cycle time to be less than 1 minute. The human operator in this case often has time to think about other things, otherwise the job does become boring, although it need not. Intermittent work should be introduced to avoid fatigue. If a work cycle is less than 1 minute then it should be mechanized if possible.
3. *'Workers must produce a complete product.'* This is rubbish! Some jobs need to be broken down into smaller parts for greater productivity and throughput. It is important to try to engineer into the assembly some aspect of quality for which the operator/group is responsible.
4. *'Everyone should plan their own work.'* This is just not practical. Planning is essential and this may involve not only the operation alone, but also material systems, tooling, and layout of the workbench. It is good to involve the operators in teamwork exercises so that they can contribute to certain aspects of planning, e.g., through quality circles, quality of work life programmes, etc.
5. *'Autonomous groups will make the foreman or supervisor redundant.'* Even if self-contained work groups are formed, the work group themselves will soon select a leader. Each of us needs leadership. Corrections done in a kind and upbuilding way are good for us and for the company.
6. *'Time study is dead.'* Many techniques are now available which are far better than time study, e.g., MOST (Maynard Operation Sequence Technique). But some time studies will be necessary to produce synthetic data for parts of the operation, such as process time related to inches cut or stitched. To know the time taken to perform operations is most essential in any form of work, otherwise proper planning cannot proceed.
7. *'A BSI replacement for work study is needed.'* The British Standard definitions have served well for many years and will continue to do so for many more. Work study has contributed much to large improvements in effective operations. Bad practice is also prevalent today. This does not invalidate the positive benefits of work study. Some change is necessary, for example the definition of standard performance as $1.2 \times$ MTM. MTM is now the internationally accepted comparison basis for times.
8. *'Individual incentives are ineffective.'* This is just not true. Properly administered individual incentives produce very good benefits in productivity. If teams or groups are working together then group schemes may be more appropriate.

There is much positive evidence that changing from individual to company-wide schemes or to daywork has reduced productivity over a few months. Incentive schemes must be reviewed regularly. Most incentive schemes if they are properly designed will last for about five years. Some schemes such as Improshare may be very relevant.

9. *'MTM is superseded.'* There are now better application systems available, e.g., MOST (Maynard Operation Sequence Technique), CAPES (Computer Aided Planning and Estimating System), etc., but they are often based on MTM. MTM is the building block for better systems of measurement.

8

Automation

8.1 Introduction

Automation has been defined as the technology concerned with the design and development of processes and systems that minimize the necessity for human intervention in their operation.

From this definition, it can be seen that there are various levels of automation, depending on the cost and complexity of the process and the amount of human intervention required. In this sense, automation has been with us for a number of years, but it is more recently with the advent of computers and microelectronics that significant developments have taken place.

Since the early 1900s there have been many machines which could have been classified as 'automatic', in the sense that the mechanisms function without intervention by a human operator; examples are the multi-spindle automatic lathe, automatic shutter control in a camera and automatic feed mechanisms for power presses.

Automatic processes have undergone three main stages in their development, as follows:

1. The first mechanisms were largely mechanical devices using the interaction of levers, cams, springs and the like; many of them were highly elaborate and ingenious.
2. A clearly recognizable new stage was reached when electromechanical systems were developed. Electrical sensing mechanisms were merged with the mechanical devices; here, solenoids and relays based on the principles of electromagnetism were used.
3. The advent of microelectronics and computers has developed the automatic process still further. Many automatic actions are now program-controlled.

Automatic control today can include a large range of options. It is often based on the automatic sensing of process variables such as position, speed, status, temperature, pressure, etc.; these provide the stimuli for automatic control actions to be performed—usually by stored program control (SPC) with outputs to the automatic control devices.

The range of applications can include the following:

1. Sequence controllers for automatic startup and shutdown of plants or processes—as in power stations for generating electricity supply.
2. Controlling the ingredients and mixing operations in batch control in the food industry.
3. Logic controllers for automatic plant as in position control, bottle filling, cap insertion and weight checking.
4. Numerically controlled machines; this usually refers to machine tools for metal removal, forming and shaping.
5. Robots in a production line, performing operations such as spot-welding of car bodies and paint spraying as part of the whole handling and storing process.
6. Automatic assembly machines—where individual parts are positioned and fastened and the assembly delivered to an output station.
7. Flexible manufacturing systems. These are systems of machine tools and other work stations integrated with the automatic transfer of components and tooling, all linked to a central computer which controls, monitors and provides information. The main application of such systems is to batched work, where changes in design and specification can be readily accommodated.

The applications just listed are distinctly different from those in office automation and business-data processing, although there may be links between them. The skills needed will be very different from those of the business data processing environment. Automation of manufacturing needs different trained personnel such as technician engineers, instrumentation and control engineers, control programmers and chartered engineers who specialize in machine tool technology, manufacturing engineering, and electronic or electrical engineering, and would act as project managers for such installations.

8.2 Low-cost automation (LCA)

Although pioneered a few years ago, as an attempt to get British industry to use many of the lower-cost units of automatic control of processes, this is still an area where many productivity gains could be made, and where worthwhile economic advantages are proven.

Such a system usually includes a wide variety of low-cost aids which can be combined for a specific end-purpose. Such aids are vibratory bowl feeders, indexing feeders, and follow-on tools linked to sensing heads, electronic relays, solenoids, or hydraulic and pneumatic circuits and control devices.

An integral part of any automation programme must be the design of products for manufacture. This includes, among other considerations, the following:

1. Using the smallest number of parts possible in an assembly. There are usually three costs associated with each part:

    (a) Making or purchasing the part.
    (b) Controlling, storage and transportation, together with unit containers for the part.
    (c) Assembling the part.

2. Making any assembly that is necessary as quick as possible. This can be achieved by the following:

    (a) Using the correct type of fastener.
    (b) Strict quality specifications and control of tolerances, fits, etc.
    (c) Employing assembly work aids such as specialized electric or pneumatic tools, work holding and stacking devices, component feed and positioning aids, functional work stations, and a balanced assembly line.
    (d) Automating sub-assemblies as much as possible. Final assembly should be quick and should not involve rejection of parts or sub-assemblies and rebuild: this should be done before final assembly.
    (e) Links with unit mechanical handling and particularly packaging requirements, palletization, track conveyors and transportation.

Value analysis and value engineering are valuable management tools to simplify designs, to eliminate unnecessary functions, or to increase the number of functions in order to achieve improved value (for instance, including components previously sold as 'extras' as an integral part of an automobile).

Part 3

Work measurement

9

Work measurement

9.1 Introduction

Work measurement, as the name implies, is the measurement of work in order to establish the *time* to perform a task. It may be measured in such a way that ineffective time is shown up which can be separated from effective time.

The *time* must be stated at a defined level of performance. Figure 9.1 shows how various performance levels may be compared. The MTM scale is

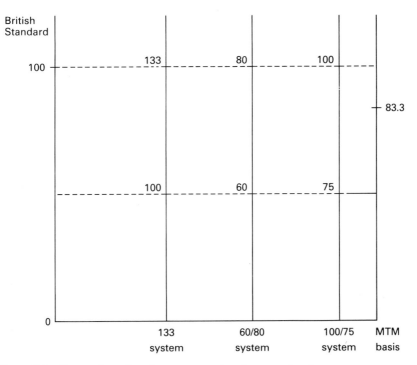

Figure 9.1 Comparison of work measurement performance levels

universally recognized as being the most consistent and most reliable basis for comparing times. It is generally agreed that MTM is equal to 83.3 on the British Standards Institution (BSI) scale of performances. MTM is generally reckoned to be the American equivalent of the British Daywork rate of working; in fact, it is slightly higher because the British equivalent is at 75 on the BSI scale shown by the dotted line above.

The situation is confusing for three reasons:

1. The ILO (International Labour Office), which incidentally has produced an excellent book called *Introduction to Work Study*, has defined normal working by virtue of the concept that an average worker would produce a performance distribution the mean of which is generally reckoned to be BSI 75. This is also taken to be the equivalent of a walking pace of 3 miles per hour; 4 miles per hour would thus be equivalent to BSI 100.
2. It is generally known that performance distributions of observed times do not behave in a way that would be similar to a normal distribution for a day worker's performance. Statistical distributions for times are usually skewed, the more favoured motion being near the standard time, which is more often repeated. Abruzzi did much research work to try to establish a relationship between rating and the measurement of skew in distributions of studied time, but generally no such relationship has been accepted.
3. Daywork performance is very rarely at 75; it is more likely to be between 50 and 60 BSI. Actual performance measures would normally include many elements of ineffective time.

The most convenient way of defining BSI 100 performance is to state that it is equivalent to $1.2 \times$ MTM. At least we all then know what we are talking about, and have a unique basis for comparison.

9.2 Purpose of time standards

In the process of setting standards it may be necessary to use work measurement to:

1. Compare the efficiency of alternative methods.
2. Balance the work of members of teams so that, as near as possible, each member has a task taking an equal time.
3. Determine the number of machines an operative can run.

Once the time standards are set they may then be used to:

1. Provide information on which the planning scheduling of production can be based, including the plant and labour requirements, for carrying out the programme of work and the utilization of available capacity.

WORK MEASUREMENT

2. Provide information on which estimates for tenders, selling prices and delivery dates can be based.
3. Set standards of machine utilization and labour performance which can be used for any of the above purposes and as a basis for incentive schemes.
4. Provide information for labour-cost control and to enable standard costs to be fixed and maintained.

It is therefore clear that work measurement provides the basic information necessary for all the activities of organizing and controlling the work of an enterprise in which the time element plays a part.

9.3 The technique of work measurement

The following are the principal techniques by which work measurement is carried out:

1. Time study.
2. Activity sampling.
3. Synthesis from standard data.
4. Predetermined motion time systems (PMTS).
5. Estimating.
6. Analytical estimating.
7. Comparative estimating.
8. Group timing technique.
9. Multiple regression analysis.
10. Computer estimating and planning system (CEPS).

A decision matrix, showing the advantages and disadvantages of various techniques of work measurement, is provided in Fig. 9.2.

STANDARD PERFORMANCE

This is the rate of output which qualified workers will naturally achieve without over-exertion, as an average over the working day or shift, provided they know how and adhere to the specified method, and provided they are motivated to apply themselves to their work. It is recommended that this is denoted by 100 on the BSI, corresponding to the production of 1 'standard hour' (SH) or 60 'standard minutes' (SM) of work in 60 minutes.

THE STANDARD TIME

The object of work measurement is to ascertain the work content of an

System	Accuracy consistency	Method orientation	Speed of application	Operator acceptance	Ease of audit or update	Study cost
Estimating	Low	None	High	Difficult to justify	Difficult	Low
Historical records	Unreliable	None	Varies	Varies with complexity	Difficult	Low
Time study	Acceptable	Low	Medium	Low	Medium	Medium
Sampling	Varies with sample size	None	Low	Medium	Difficult	High
Basic predetermined time systems: □ MTM-1 □ BMT (basic motion time study) □ Work factor □ Modapie	High	High	Low	Medium	Easy	Medium to high
High-level predetermined Time systems: □ MTM-2 and -3 □ MTM-	High	High	Medium	Medium to high	Easy	Low to medium
Fourth-generation: □ MOST □ Computerized (CEPS)	High	Medium	High	Medium to high	Easy	Low

Figure 9.2 Measurement systems—decisions matrix

operation or task. The methods used to achieve this have changed significantly over the past few years, although the components of a 'standard' time have remained constant in outline. These are:

1. *Basic work content*   The time to actually perform the operation or task at a defined level of performance. The universally recognized basis is the MTM method of establishing standard data, which is the most consistent and reliable of all the methods.

    The British Standard for basic work content, called 'basic minutes' or 'basic hours', is evaluated at $1.2 \times$ MTM. When time study is used a technique called 'rating' is used to convert actual minutes observed to basic minutes of work content.

2. *Contingency allowance*   This is normally a small percentage allowance on the basic work content, to allow for infrequent work which is normally carried out but which has not been allowed in the basic work content. This allowance should always be established by actual observations over a period of time, and would be different for each organization and type of task.

3. *Relaxation allowance*   This is a further percentage for two different requirements:

    (a) *Personal needs*   This normally includes allowances for going to the toilet, for 10 minutes of refreshments each morning and each afternoon, and for any stoppages which occur throughout the working day of less than two minutes' duration. The refreshment breaks need not be specified periods, but could be taken at any time, depending on the local agreement.

    (b) *Fatigue allowances*   These are relaxation allowances to overcome physiological fatigue because of special needs of the task in hand; for example, working in a hot atmosphere or constantly lifting weights. Many of these factors ought to be designed out of the task. Most practitioners would use tables prepared by the ILO for this purpose, although many countries have published their own standards for this component.

When the three components described above are added together, the result is usually described as the 'standard time' and in British Standard terminology is stated as 'standard minutes' (SM) and 'standard hours' (SH). Any allowances in addition to the above should not be described as standard time. A bad practice often used in industry is to increase the standard time by a policy allowance; such a standard should be called an 'allowed time' and not a 'standard time'.

INDUSTRIAL ENGINEERING

9.4 Predetermined motion time systems

These are work measurement techniques whereby times established for the human motions (classified according to the nature of the motion and conditions under which it is made) are used to build up the time for a job at a defined level of performance.

The first generation of these techniques were very detailed and the basic movements were broken into very minute motion patterns (MTM-1, Work Factor). In the second generation it was found that in many applications the small motion patterns were continually being repeated, giving rise to synthesized data for broader elements. For instance, in MTM-2, single basic MTM motions were derived which were combinations of MTM-1 motions.

The advantages of MTM-2 over MTM-1 are:

1. Consistency between analyses.
2. Speed of handling.
3. Universal nomenclature.
4. Ease of understanding.
5. Descriptiveness of the method.
6. Capacity to be combined with other MTM data.
7. Specification in relation to speed of accuracy and accuracy of results.

MTM-2 is intended for operations longer than 0.75 minutes, with an accuracy of ± 5 per cent at time-cycles of a single minute. It is usually reckoned to be four times faster to apply than MTM-1.

MTM-3, which uses larger groups of motions than MTM-2, was developed in Sweden, particularly for maintenance and non-repetitive work. It provides a 95 per cent confidence level of ± 5 per cent accuracy at time-cycles of 10 minutes' duration and ± 10 per cent accuracy at time-cycles of 2.5 minutes' duration. It is reckoned to be three times faster to apply than MTM-2.

Other developments used in Britain include TDA (tape data) and data block approach. PMTS systems have been developed particularly for specific applications. For example:

– UMS   universal maintenance standards.
– CWD   clerical work data

THIRD-GENERATION TECHNIQUES

The most recently developed technique of PMTS, i.e., MOST, allocates a series of times for a whole sequence of motions. MOST is based on what are called 'sequence models' for different kinds of manual activity. The following are examples:

WORK MEASUREMENT

Activity	Sequence model	Sub-activities	
1. General move	ABGABPA	A	action distance
		B	body motion
		G	gain control
		P	place
2. Controlled move	ABGMXIA	M	move controlled
		X	process time
		I	align
3. Tool use	ABGABP ABPA	F	fasten
		L	loosen
		M	measure
		R	record
		S	surface treat
		T	think

In addition to the above there are also categories for heavy work involving:

1. Move with jib crane.
2. Move with bridge crane.
3. Move with wheeled truck (manual and powered).

9.5 Methods of determining the basic work content

There are many different methods in use for determining the work content of a task. The greatest need is to establish a fast but reliable method for work measurement by way of:

- Accuracy
- Consistency
- Simple data development
- Economy
- Ease of use and updating
- Methods orientation

THE NEED FOR ACCURACY

Factors of accuracy

Four factors contribute to the accuracy of work measurement:

The level of accuracy desired

This is normally considered by practitioners to be ± 5 per cent.

Table 9.1 Balancing times for MTM-based systems

Work measurement technique	Balancing time	
	TMU	seconds
MTM-1	600	21.6
MTM-2	1 600	57.6
MOST	3 200	115.2
MTM-3	15 000	540

Note: TMU (time measured unit) = 0.00001 h, or 0.00006 min, or 0.036 s.

The balancing time

This is the period over which the desired level of accuracy must be attained, for example eight hours or forty hours.

The balancing effect is defined as 'the levelling out of individual deviations for a smaller total deviation'. See Table 9.1.

Degree of repetitiveness of activity

If a job repeats itself frequently, there needs to be a greater consistency of the value—measured by the deviation of the mean value and by the standard deviation of the scatter of the values. If the same job repeats itself only occasionally, as for instance with maintenance work, then the operator will not have time for the learning curve to take effect.

Duration of the activity being considered

If the proportion of total working time taken by an activity is high then there is a need for greater accuracy. The spread of values for a short-cycle assembly job should be much less than for a maintenance job, where the total allowable deviation should be greater—sometimes a maintenance job will go very easily and at other times the same job will have stiff bolts, say, or bearings which have seized hard.

The theory of added variances

This theory can adequately cover the four factors just listed; it is expressed as follows:

$$r_t = R_T \sqrt{\frac{T}{n \times t}}$$

WORK MEASUREMENT

where

r_t = the measured activity's allowed deviation (per cent)
R_T = the required total allowed deviation (per cent)
T = the total time (balancing time) (hours)
n = the activity's frequency of occurrence during T
t = the activity's measured time (hours)

Tables have been designed to take into account each of these factors when applying standard times. One of the faults common to many practitioners is that they often set standards with more implied precision than is the actual case.

THE NEED FOR CONSISTENCY

Many techniques of work measurement are prone to analyst errors. This is particularly true of the more detailed PMTS systems such as MTM-1; it is also true of time study. Attempts have been made to reduce inconsistencies through the use of synthetics and databases of elemental times, or through rating clinics; but it is still surprising how varied are the results of work measurement programmes within companies. The new techniques such as MOST and computerized work measurement are much more consistent.

THE NEED FOR SIMPLE DATA DEVELOPMENT

The third generation of work measurement techniques such as MOST offer simplicity of application. Numerous comparison studies have shown that they are as accurate as other far more complicated systems.

In terms of compactness, the more detailed the analysis necessary, the greater the number of the pages of documentation and the greater the amount of practitioner skills needed to maintain consistency. The example given in Table 9.2 is typical of some PMTS techniques.

Table 9.2 Example of a PMTS technique: assembly of a carburettor

Work measurement technique	No. of pages of documentation used	Standard time, TMU
MTM-1	16	4402
MTM-2	10	4446
MTM-3	8	4950
MOST	1	4530
Time study	4	varied

Table 9.3 Application speeds of MOST and older systems compared

Work measurement technique	Total of TMUs assigned per analyst hour
MTM-1	300
MTM-2	1 000
MTM-3	3 000
MOST	12 000

THE NEED FOR ECONOMY

The latest techniques of work study—computerized systems or MOST—are much faster to apply than the older techniques such as time study. In terms of PMTS, Table 9.3 gives an indication of application speeds and hence costs. MOST is considered to be at least three times as fast as time study to apply.

As far as computerized work measurements systems are concerned, this will depend on the application considered. Generally, they are faster to apply—needing only an applicator rather than a standards development analyst.

The main economy of computerized systems is that they can combine standards application, process planning, cost estimating, and typing of documents together. Some companies claim that computerized systems take between 10 and 15 per cent of the total time for the equivalent manual systems.

THE NEED FOR EASE OF USE AND UPDATING

One of the problems with time study is that methods can change without the analyst being aware that anything has happened, only that the times are getting loose. This may be because the application was incorrect in the first place, the learning curve effect having occurred to loosen times. It may be that genuine method changes have been effected, but that elemental times and descriptions are inadequate to detect the difference.

Two separate problems exist alongside this:

1. Wage drift is a common occurrence.
2. Maintaining accurate time standards is also very difficult, because of shop floor and union attitudes prevalent in the UK.

Any systems of work measurement in use should be easily updated, and should be readily understood in principle by operatives, so that they are convinced of the fairness and consistency of the results.

Table 9.4 MOST techniques and their applications

Technique	Index base	Applied to	Previous technique
Mini-MOST	1	Short-cycle operations	MTM-1
MOST	10	Ordinary operations	MTM-2 Time-study
Maxi-MOST	100	Non-repetitive work, e.g., maintenance	MTM-3 UMS
Clerical MOST	10	Clerical operations	MTM-C

Table 9.4 lists the various applications of MOST techniques. There are different versions of MOST, now a complete system of work measurement which has virtually rendered MTM and time study obsolete in terms of applicability, though not in terms of practice. The United Kingdom has so far been very slow to adapt to this method, although it has been used in Sweden, the United States, Western Europe and even in the Eastern bloc, e.g., Hungary.

9.6 MOST sequence models

Until recently MOST has been used exclusively on an individual basis by clients of H. B. Maynard, the international consultancy company. It was developed in Sweden between 1967 and 1972, and then further developed in the United States after 1974. It has now been used in many parts of the world. In England perhaps the most notable installation is at J. C. Excavators Ltd of Uttoxeter, which has also implemented the computerized MOST version.

Work measurement has traditionally broken work down into discrete, very small sub-divisions of movement, called 'basic motions'. This may be necessary for short-cycle work, and indeed has been found to be a convenient way of work design and measurement for highly repetitive work. Longer operations, however, either have undergone this particular way of analysis, with the result of uneconomic and complicated studies and calculations, or have been left unmeasured. This especially applies to work such as maintenance, jobbing manufacturing, and non-repetitive work.

MOST was developed using an entirely different approach to the problem of analysis. The first concept was to ask what work is. For those trained in science and technical subjects, the answer is 'exerting energy' or 'the displacement of a mass', and is calculated by the formula, work = force × distance. This can be applied to any form of physical work; thinking or conceptual work does not quite fit into the scheme. Simply stated,

WORK IS THE MOVEMENT OF OBJECTS

MOST is based on that premise. It is based on the whole sequence of events when we move objects. These sequences are described using *models* to represent basic types of sequences that can be found in practice.

THE THREE BASIC MODELS OF SEQUENCE

These are:

1. *General move*   The movement of objects without any form of restriction (except air resistance which can be neglected).
2. *Controlled move*   The measurement of objects when there is some form of restriction (e.g., slide, hinge, stop), i.e., which restricts free movement in some way.
3. *Tool use*   When the movement involves holding something else, e.g., a tool such as a pencil, spanner, rag, etc.

General move

The general move has the following sequence, involving four different sub-activities:

$$\underbrace{A\ B\ G}_{\text{Get}} \quad \underbrace{A\ B\ P}_{\text{Place}} \quad \underbrace{A}_{\text{Return}}$$

A is the action distance (horizontal movement).
B is body motion (vertical movement mainly).
G is gain control.
P is place.

The time-related values for each of these sub-activities are obtained from an application data card. A copy of part of one of these is provided in Fig. 9.3.

With the majority of PTMS systems an element of time is selected for the defined basic motion. With MOST a different basis is used. It starts by defining intervals of time—these get larger and larger and are based on a standard balancing time (see page 64). It then selects the motion sub-activities which fall into each time-interval.

These time-interval groups are also based on the following premises:

1. The mean value for each time-interval will be a whole number and a multiple of 10.

WORK MEASUREMENT

GENERAL MOVE	ABGABPA			
	A	**B**	**G**	**P**
INDEX	ACTION DISTANCE	BODY MOTION	GAIN CONTROL	PLACE
0	≤2 in ≤5 cm			Hold Toss
1	Within Reach		Light Objects Light Objects Simo	Lay Aside Loose Fit
3	1-2 Steps	Bend and Arise 50% occ	Non Simo Heavy or Bulky Blind or Obstructed Disengage Interlocked Collect	Adjustments Light Pressure Double
6	3-4 Steps	Bend and Arise		Care or Precision Heavy Pressure Blind or Obstructed Intermediate Moves
10	5-7 Steps	Sit or Stand		
16	8-10 Steps	Through Door Climb On or Off		

Figure 9.3 Application data card (Courtesy H. B. Maynard & Co. Ltd)

2. The time-intervals will cover a continuous time-scale, with no gaps and limited overlap.

Figure 9.4 shows the principle adopted. The values on the application

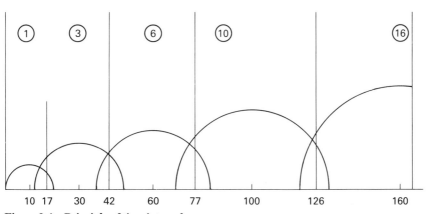

Figure 9.4 Principle of time interval groups

card are those shown circled. The numbers will need to be multiplied by 10 to obtain TMUs. A time-indexed general move sequence will appear as follows:

```
       A  B  G  A  B  P  A
      10  6  3  1  0  1  0
```

A
  10   Walk six steps to location.
B
  6   Bend and arise.
G
  3   Gain control group collection of items in hand.
A
  1   Move within releasing distance.
B
  0   No bend.
P
  1   Place down collection of objects.
A
  0   No return.

The TMU time value for the sequence would be as follows:

$$(10+6+3+1+0+1+0) \times 10 = \underline{210 \text{ TMU}}$$

Controlled move

The sequence model for the controlled move is:

```
       A  B  G  M  X  I  A
```

where A, B and G are the same sub-activities for the general move and

    M = move controlled (controlling levers, cranks, wheels and switches)
    X = the process time (also based on time indexes)
    I = align (following the process time)

Tool use

The sequence model for tool use is

```
       A  B  G  A  B  P  . . . . . .  A  B  P  A
```

Sub-activities codified as follows are inserted into the sequence model where the dotted line is shown:

F = fasten with a tool R = record
L = loosen   S = surface treat
M = measure   T = think

Other sequence models used in MOST are

- Move with a jib crane.
- Move with a bridge crane.
- Move with a wheeled truck (e.g., hand, powered, forklift, etc.).

9.7 Training in MOST

Training courses in MOST and the award of practitioner certificates are controlled in the United Kingdom by H. B. Maynard & Co., London. Some companies in the United Kingdom are licensed by H. B. Maynard to run training courses for their own staff.

One of the authors has been a senior consultant applying the technique with Maynard overseas. During this time, the work measurement programme for a factory unit of just under a thousand people was completed in 12 weeks with a team of people recruited from production engineering and the shop floor. Prior to this, they were trained for three weeks (theoretical and practical). A further period was used in writing up the work management manuals. With normal time study this programme would have been at least twice to three times as long, and would have needed a much longer period of training.

9.8 MOST calculation sheet

An example of a MOST calculation sheet is provided by Fig. 9.5.

9.9 Further refinements to MOST

These were introduced during 1982, namely Mini-MOST, Maxi-MOST and Clerical MOST.

MINI-MOST

To be used when cycle times are very short and the motions pattern is identical with each cycle. Mini-MOST retains both the general and controlled move sequence models, but uses an index multiplier of × 1 (instead of × 10 for MOST).

Mini-MOST has its own data card and parameter definitions.

INDUSTRIAL ENGINEERING

MOST-calculation

Code	0 0 7 8 3 62 2 0 1
Date	11.3.73
Area: CAR PRODUCTION	Sign. LS
	Page 1 / 1

Activity: ASSEMBLY - FAN INSTALLATION

No	Method	No	Sequence model	Fr	Time
1	Fasten 4 screws with torque spanner	2	$A_1 B_0 G_3 A_0 B_0 P_0 A_0$	2	80
		3	$A_3 B_0 G_3 A_3 B_0 P_1 A_0$		100
2	Detach 2 plastic covers	4	$A_3 B_3 G_3 A_3 B_0 P_3 A_0$	1/10	15
		6	$A_0 B_0 G_0 A_0 B_0 P_1 A_0$		20
3	Detach 3rd plastic cover	7	$A_1 B_0 G_1 A_1 B_0 P_1 A_0$		40
		8	$A_1 B_0 G_1 A_1 B_0 P_1 A_0$		160
4	Get 10 fans	9	$A_1 B_0 G_3 A_1 B_0 P_6 A_0$		110
		10	$A_1 B_0 G_1 A_1 B_0 P_3 A_0$		240
5	Remove 2 screws from fan cover	11	$A_1 B_0 G_1 A_1 B_0 P_0 A_0$		30
		14	$A_0 B_0 G_0 A_1 B_0 P_1 A_0$		20
6	Aside screws		A B G A B P A		
			A B G A B P A		
7	Assemble space bar		A B G A B P A		
			A B G A B P A		
8	Assemble washer		A B G A B P A		
			A B G A B P A		
9	Assemble fan		A B G A B P A		
			A B G A B P A		
10	Assemble screw		A B G A B P A		
11	Move to machine	13	$A_0 B_0 G_1 M_1 X_0 I_0 A_0$		20
		15	$A_0 B_0 G_1 M_1 X_0 I_0 A_0$		20
12	Re-set machine		A B G M X I A		
			A B G M X I A		
13	Fasten 4 screws for fan		A B G M X I A		
			A B G M X I A		
14	Move to machine	1	$A_1 B_0 G_1 A_0 B_0 (P_3 A_1 F_3)$ $A_1 B_0 P_1 A_0$	(4)	320
		5	$A_1 B_0 G_1 A_0 B_0 (P_3 A_1 F_3)$ $A_1 B_0 P_1 A_0$	(2)	180
15	Re-set machine	13	$A_0 B_0 G_0 A_1 B_0 (P_3 A_1 F_3)$ $A_1 B_0 P_1 A_0$	(4)	300
		16	$A_1 B_0 G_1 A_1 B_0 (P_3 A_0 F_3)$ $A_1 B_0 P_1 A_0$	(4)	290
16	Fasten 4 screws with torque spanner		A B G A B P A B P A		
			A B G A B P A B P A		
			A B G A B P A B P A		
			A B G A B P A B P A		
			A B G A B P A B P A		
			A B G A B P A B P A		
			A B G A B P A B P A		
			A B G A B P A B P A		

TIME = 1.17   XXX/MINUTES   1,945

Figure 9.5 MOST calculation sheet

WORK MEASUREMENT

MOST-calculation

Code: 0 0 7 7 0 2 5 3 0 1
Date: 5.1.75
Sign.: KZ
Page: 1 / 1

Area: HOSPITAL CANTEEN

Activity: Load Food Trolley - Trayline

No	Method	No	Sequence model	Fr	Time
1	Get food trolley and open doors (A)	3	$A_3\ B_0\ G_1\ A_1\ B_0\ P_3\ A_0$	1/10	8
2	Close doors (B)	4	$A_3\ B_0\ G_1\ A_3\ B_0\ P_1\ A_0$		80
3	Place identification label on trolley	5	$A_6\ B_0\ G_1\ A_6\ B_3\ P_3\ A_1$		200
4	Place banquet cover on plate		A B G A B P A		
5	Load tray in trolley (max. 16 trays, average 10)		A B G A B P A		
6	Relocate trolley and push aside		A B G A B P A		
		1	$A_6\ B_0\ G_1\ M_3\ X_0\ I_0\ A_6$	1/10	16
		2	$A_1\ B_0\ G_1\ M_3\ X_0\ I_0\ A_0$	4/10	20
		6	$A_1\ B_0\ G_1\ M_3\ X_0\ I_0\ A_0$	2/10	10
			A B G M X I A		
			A B G A B P A B P A		
			A B G A B P A B P A		
			A B G A B P A B P A		
			A B G A B P A B P A		
			A B G A B P A B P A		
			A B G A B P A B P A		
			A B G A B P A B P A		
			A B G A B P A B P A		
			A B G A B P A B P A		
			A B G A B P A B P A		
			A B G A B P A B P A		

TIME = .2 TMU/MINUTES   334

Figure 9.5 (*continued*)

INDUSTRIAL ENGINEERING

MAXI-MOST

Maxi-MOST is used for long cycles of work. Three new sequence models are used:

- ABP   Part handling
- ABT   Tool use
- ABM   Machine handling

CLERICAL MOST

This is similar to MOST, with an index multiplier of × 10, with the same general move and controlled move, and with an equipment-use model. Specific time-ranges are employed for paper handling, typewriter manipulation, electric calculators, electronic calculators, read and record, and various tool uses.

9.10 Need for methods orientation

One of the biggest drawbacks with time study is that, being based on 'element descriptions', there is very limited methods orientation. With MTM systems, by designing the method one is automatically obtaining the time. Time and method are related—if you walk twice as far it takes double the time, if you pick up an item from twice the distance it takes double the time, and so on. The effect of method on time is directly seen with PMTS, but this is not so with time study.

10

Computerized systems of estimating, planning and standard data

10.1 Introduction

These are systems of data building based on computer programs and on storage of basic data in disk files. It is usual to link parts of the system together into a chain of activities combining methods engineering, process planning and selection of process-time variables (such as for machine shops, sewing machines or chemical plant variables). Data can be built up into various levels of precision and application, as follows:

Level	Examples
3rd	for estimating and pre-planning (fairly coarse data)
2nd	for factory loading, production planning group bonus scheme (combining averages of data, etc.)
1st	elemental time standards or for individual incentive scheme (based on MTM-2 or time standards)
Elementary	back-up data—elementary MTM-1 patterns and minute process-time data

As the nature of the data approaches a higher level, so the precision becomes less, but the application time becomes faster.

The various types of application for computers in work measurement are as follows:

1. *Computerized time study* This is usually the working up of a stopwatch-based time study comprising normalizing, selection of basic minutes from a time study sample, selection and addition of allowances.

INDUSTRIAL ENGINEERING

Table 10.1 PC software packages

Name	Operating system	Min main memory	Floppy disk	Hard disk	Supplier
FAST	PC-DOS	128K	1 × 320K	5 mb	Tectime
TEL WORKSTUDY	PC-DOS	128K	2 × 320K	No	Telford Management
TEL ACTIVITY SAMPLING	PC-DOS	128K	2 × 320K	No	Telford Management
TIMESTUDY	PC-DOS	128K	2 × 320K	5 mb	Mercator
WORKSTUDY	PC-DOS CP/MS6	128K	2 × 320K	No	Tectime

The output is a list of element descriptions and basic minutes, time study summary, and standard time for the task.

2. *Computerized synthetics*   The programs are primarily for the storage, manipulation, and printing of standard data, usually obtained from time study-based synthetics or MTM patterns. Synthetics would normally be stored on disk in the form of tables or formulae.

The output is typically:

(a) Element description and standard times.
(b) Process planning documents.
(c) Linked to graphics mode, display of tables and formulae in line chart or histogram related to the process variable.

3. *Computerized estimating and planning systems (CEPS)*   Here the standard data is held on disk storage for a wide range of industries and can be extracted to provide:

(a) Operation breakdown, including speeds and feeds etc.
(b) Operation description and standard time.
(c) Process planning documents.
(d) Cost estimates.

4. *Integrated standard data*   This can be CEPS where this is linked to the integrated system for manufacturing. The CEPS is in this case a module in a CIM (computer integrated manufacturing system) or MRP II system (Manufacturing Resource Planning).

COMPUTERIZED SYSTEMS

The chain of linkage is typically

$$CAD > CPE > CPC > CAM$$

10.2 Computerized time study

Table 10.1 lists the packages entered in the IBM software directory for the PC. Prices range from £750 to £2000.

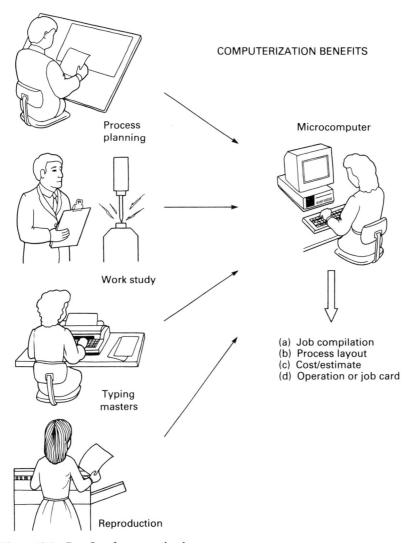

Figure 10.1 Benefits of computerization

10.3 Computerized estimating and planning systems (CEPS)

The main advantages of a CEPS system is that it combines a number of clerical and/or technical operations previously done by separate people and often separate departments, namely:

- process planning/planning engineer
- work study officer/engineer
- typist (masters)
- reproduction (of documentation)

See Fig. 10.1.

CEPS SOFTWARE

The main companies who market software for CEPS are:

Name of package	Main supplier
LOCAM	Logan Associates Ltd
	PE Consultants Ltd
CAPES and	Methods Workshop Ltd
SUPERCAPES	Inbucon Productivity Services Ltd
	Scicon Ltd
SOFIE 2	Organization Development Ltd
CEEQUEL	Scott Grant Ltd
CIMPLEST	Westborough Computer Services Ltd

System structure

A typical system structure is shown in Fig. 10.2. Its features are:

1. Elimination of manual searches through charts and reference materials.
2. Data banks of machine and material specifications which ensure reliability.
3. Multiple reports which aid in the efficient running of operations.
4. Calculation of time and costs which are computer accurate.
5. Menu-driven system which is easy to learn, simple to use.
6. Automatic storage of estimate for future reference.
7. Data banks which, once created, can be updated by user at any time.
8. Reduction of estimating time and increase in efficiency.

Standard data

Types of data available for such systems now cover a variety of operations.

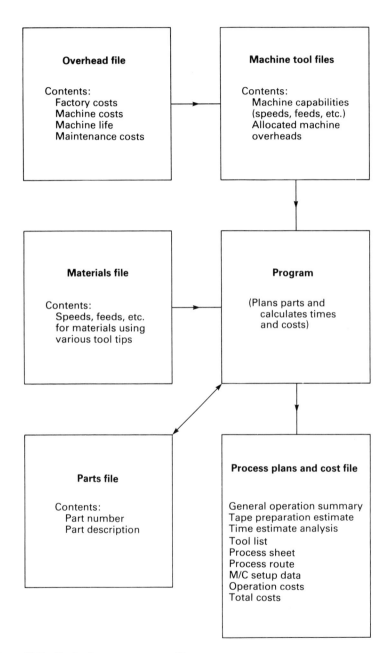

Figure 10.2 Typical system structure diagram

INDUSTRIAL ENGINEERING

Below are listed a number of them. They are normally supplied on a floppy disk for each module but can be incorporated onto hard disks.

Machine shops
- Centre lathes
- Capstan lathes
- Turret lathes
- Milling
- Radial drill
- Gang pillar drills
- Internal grinding
- External grinding
- Surface grinding
- Horizontal boring
- Vertical boring
- Gear grinding
- Shaping
- Deburring

Toolroom
- Detailed estimating data
- 'Quick' manpower planning data

Assembly
- Long-cycle-time assembly for large products
- Light benchwork assembly at MTM level using core data MTM-2
- Electrical/electronic assembly

Fabrication
- Cut-off saws
- Bandsaws
- Guillotines
- Flame cutting
- Marking out
- Setting and tacking
- Welding
- Spot welding
- Power nibbler/sheer
- Press work—Hand press
  —Power presses
  —Press brake
  —Peddinghaous punch
- Hand rolls
- Power rolls

Other areas covered
These include:
- Dynamic balancing
- Static balancing
- Fettling
- Plant maintenance trades (all trades)
- Vehicle maintenance
- Clerical—all areas
- Manpower planning— retail trades

A typical operation data sheet for a CAPES system is provided in Fig. 10.3. Operation process planning documents are also produced, together with individual data sheets, operation costings, and tape preparation estimates (Fig. 10.4).

10.4 Other computerized work measurement systems

These can vary, but examples are the following:

1. *REAP 3* (from P-E Consultants Ltd)   This employs statistical indexed

```
+-----------------------------------------------------------------------------+
|                    W E S T B O R O U G H                                    |
|                    TIME  ESTIMATE  ANALYSIS                                 |
+-----------------------------------------------------------------------------+
| CUSTOMER :-   PRECISION ENGINEERING COMPANY                                 |
+-----------------------------------------------------------------------------+
| ESTIMATED BY :-   S.C.CARMICHAEL        | DATE :-  08-16-85                 |
+-----------------------------------------------------------------------------+
|     PART NUMBER      | ISSUE |            DESCRIPTION                       |
|     H801559          |   A   |            MOUNTING BLOCK                    |
+-----------------------------------------------------------------------------+
|        COMPONENT MATERIAL            |         BLANK SIZE                   |
|        MANGANESE STEEL               | 130.00 mm  x  130.00 mm  x  25.00 mm |
+-----------------------------------------------------------------------------+
| OPERATION : OP10              | MACHINE : YAMAZAKI V20                      |
+-----------------------------------------------------------------------------+
```

SEQ. NO.	OPERATION TYPE	No. OF CUTS	CUT DEPTH (MM)	CUTTING SPEED (M/MIN)	SPINDLE SPEED (RPM)	FEED RATE (MM/MIN)	FEED RATE (MM/REV)	LENGTH OF CUT (MM)	CUTTING TIME (MIN)
	TOOL CHANGE								0.15
1	DRILLING	3	8.333	15	592	122.520	-	250.00	2.50
DRILL, 10 HOLES , 7.5 MM.DIA. , 25 MM.DEEP (HIGH SPEED STEEL,MAX SURFACE SPEED)									
	TOOL CHANGE								0.15
2	DRILLING	4	6.250	15	630	130.570	-	50.00	0.48
DRILL, 2 HOLES , 7.0 MM.DIA. , 25 MM.DEEP (HIGH SPEED STEEL,MAX SURFACE SPEED)									
	TOOL CHANGE								0.15
3	REAMING	1	25.000	15	323	131.270	-	50.00	0.40
REAM, 2 HOLES , 7.5 MM.DIA. , 25 MM.DEEP (HIGH SPEED STEEL,MAX SURFACE SPEED)									
	TOOL CHANGE								0.15
4	END MILLING	1	15.000	59	756	154.220	-	360.00	9.32
4xEND MILL , 80MM.LENGTH , 20MM.WIDTH , 15MM.DEPTH , 4 TEETH. 25 MM.DIA. (CARBIDE,ROUGH MILLING)									
4	END MILLING	1	25.000	59	756	154.220	-	360.00	9.32
4xEND MILL , 50MM.LENGTH , 20MM.WIDTH , 25MM.DEPTH , 4 TEETH. 25 MM.DIA. (CARBIDE,ROUGH MILLING)									
	TOOL CHANGE								0.15
5	TAP DRILLING	1	18.000	15	538	81.990	-	72.00	1.53
TAP DRILL, 4 HOLES ,(M10) 9 MM.DIA. , 18 MM.DEEP (HIGH SPEED STEEL,MAX SURFACE SPEED)									
	TOOL CHANGE								0.15
6	TAPPING	2	10.000	3	98	98.000	-	80.00	0.85
TAP, 4 HOLES , M10(PITCH 1mm) , 10 MM.DEEP (HIGH SPEED STEEL)									
	TOOL CHANGE								0.15
7	BORING	2	1.000	15	238	35.700	-	44.00	1.23
BORE , HEAVY ROUGHING , CARBIDE COATED GRADE P35 (GC135) , NOSE RADIUS .8MM, 35MM.DIA., 7.5MM.DEPTH									
	TOOL CHANGE								0.15
8	BORING	2	1.000	15	136	20.400	-	24.00	1.18
BORE , HEAVY ROUGHING , CARBIDE COATED GRADE P35 (GC135) , NOSE RADIUS .8MM, 45MM.DIA., 5MM.DEPTH									

```
              SET-UP TIME                  =   110.50   MIN.

              TOTAL MACHINING TIME         =    26.81   MIN.
              TOTAL TOOL CHANGE TIME       =     1.20   MIN.
              LOAD/UNLOAD                  =     4.50   MIN.
              MANIPULATION TIME            =    10.50   MIN.
              TOTAL CYCLE TIME             =    43.01   MIN.

              BATCH SIZE                   =      10    PARTS

              BATCH CYCLE TIME             =     9.01   HRS.
```

Figure 10.3 Operation data sheet for a CAPES system

INDUSTRIAL ENGINEERING

**** INBUCON PRODUCTIVITY SERVICES ****

PART/DRG. NO.	NEC1	OP. NUMBER	40
DESCRIPTION	DEMONSTRATION	RA-CYCLE	12
ANALYST	ARP	RA-SET	14
DATE	27/9/82	M/C TYPE	ROD WELD
MATERIAL	STEEL UP TO 35 TONS	DATA CENTRE	WELDING

OP. DESC   WELD 3 PARTS TOGETHER

*** SUMMARY OF TIMES ***

TOTAL BASIC MINUTES   TOTAL STANDARD MINUTES

SET TIME	10.31		
MANIP TIME	19.03		
CUT TIME		MANIP + CUT	31.33
MANIP + CUT	27.97	SET TIME	11.75

OPERATION DETAILS

SETTING ELEMENTS

ELEMENT DESC	FREQ
CONSTANT PER SET-MONOR	1
SHOP GRD LEVEL/BENCH	1

NO.	DESCRIPTION	FREQ	BMS
1	LOAD TO JIG		
2	LOCATE IN JIG UPTO 5LB	3	.43
3	SECURE-SCREW CLAMP	6	1.04
4	LOCATE IN JIG UPTO 5LB	5	.72
5	TACK TO HOLD		
6	STEEL PER TACK-6G ROD	24	3.12
7	WELD COMPLETE		
8	ROD 6G 5.16IN FILLET/IN	45	13.5
9	REMOVE FROM JIG		
10	ASIDE FROM JIG-UPTO 15LB	1	.22

Figure 10.4 Operation process planning document. Copyright © Methods Workshop Limited, 1982

time-values related to a motion code. It uses a programmable calculator and produce time-values and estimates.
2. *REACT* (from P-E Consultants Ltd) This uses simplified data for estimating purposes only and is available on microcomputer.
   Output examples are:
   (a) Material list; assembly listing
   (b) Routing information

(c) Estimating
(d) Made in/bought out index
(e) Shop loading; cost files

3. *Computer MOST* (from H. B. Maynard and Co. Ltd) This is a more sophisticated computerized system and is used on either a minicomputer or a microcomputer. The data base uses MOST sequence models, with the following types of output:

(a) Workplace layout.
(b) Method description and standard time.
(c) Allocation of manual and process time.
(d) Line balancing program.

There are many other packages also available, such as:

– Activity sampling
– Statistical analysis
– Multiple regression analysis

10.5 Robot time and motion

There are many robot simulation packages appearing on the market. For instance, McDonnell–Douglas, IBM, Geisco, and Ingersoll Engineers have all developed these. A particular system developed for elements of motion and time for robots was by Y. N. Shimon of the School of Industrial Engineering at Purdue University. The questions he asked were:

1. Can and should robots be preferred to human operators or hard automation?
2. If so which robot models should be selected, and how should they operate?

From these questions and as a result of his research a system called 'Robot Time and Motion' (RTM) was developed.

10.6 Clerical work measurement (CWM)

Clerical work measurement has been in use as a technique for many years, with varied success. This does not invalidate the technique, but perhaps more the attitudes and environment under which it has been expected to run.

During the past decade there has been an enourmous increase in the number of office workers employed. The numbers are still increasing, even with all the attempts at improving efficiency through business systems analysis, organization and methods, modern office equipment aids, and computerization. There is, however, a vital need for controlling indirect costs in any organization because of the following factors:

1. The need for organizations to become more competitive or provide a better level of service.
2. The large increase in the costs of employing people—not just their salaries, but also the overheads such as national insurance, pensions, building space, telephones, heating, lighting and postage.
3. Changes in the social structure of our population, namely:

    – Increase in school-leaving age
    – Shorter working week
    – Longer holidays
    – Earlier retirement
    – Productivity and union agreements

Most of the larger organizations in the United States and the United Kingdom have introduced techniques for improving office efficiency involving CWM. These include banks, insurance offices, building societies, health administrations, local authorities and manufacturing companies.

The techniques for clerical work measurement are many; they include:

CWD	– clerical work data
CWIP	– clerical work improvement programme
GCA	– group capacity assessment
MCD	– master clerical data
MMD	– milli minute data
SIS	– short interval scheduling
VFP	– variable factor programming
MTM-C	– clerical version of MTM
Clerical MOST	– clerical version of MOST

A brief summary of Clerical MOST has already been included in this book (page 74).

The procedure for CWM normally includes the following:

1. Analysis of the work performed

    (a) Flow charting and measurement (O and M, CWM).
    (b) Developing new methods (O and M).
    (c) Completion of standard times (CWM).

2. Measurement of work volumes

    (a) Defining types of work.
    (b) Actual quantity and frequency of different types of work.
    (c) Maximum quantity and frequency of work.

COMPUTERIZED SYSTEMS

3. Development of work targets and manning levels
   Calculations from (a) and (b).
4. Design and implementation of control systems.
5. Measurement and control of results.

There is no doubt that good systems of administration, linking manual methods to high technology, are necessary in large offices in order to control the high costs of office work. Quality and service levels within the office environment need also to be included in business analysis studies.

10.7 Other techniques of work measurement

There are a wide variety of work measurement techniques and methods in use today. Some of these are special tables related to the type of industry being studied, e.g., garment manufacture (laying up, cutting, stitching, etc.), stocking manufacture, weaving, fully fashioned woollen operations, car servicing and maintenance, hotel work and coal mining.

There are also some which are specialized techniques applicable in most industries.

TIME SLOTTING/BENCHMARKING

This is defined by BSI as follows:

> A work measurement technique in which the time for a job is evaluated by comparing it with the work in a series of other jobs—benchmarks—the work content of which has been measured. The arrangement of jobs into broad bands of time is referred to as 'slotting'.

It is important that the benchmarks are set accurately, and that they are selected from a typical range and variety of operations within the plant. This type of system has been used in maintenance, large-ship repair work and construction.

MULTIPLE REGRESSION ANALYSIS (MRA)

When two or more variables are independent, there is no relationship between them. When two or more are dependent then we usually have a problem: is there a causal relationship between them, and, if there is, what is that relationship in mathematical terms?

When there is a variable whose value is influenced by another, this is called the *dependent* variable, denoted by y. The variable that exerts the influence is called the *independent* variable, usually denoted by x if there is only one, by other letters if there are multiple ones.

When a statistical relationship is established, we state that there is a *correlation* between the variables.

When there are only two variables, a solution is normally found by what is called the 'method of least squares' and from this we can draw a regression line accurately to the formula

$$y = mx + c$$

where
  m is the slope of the line
  c is the intercept

We can also establish a goodness of fit, between the variables and the regression line, known as the coefficient of determination R:

$$R^2 = 1 - \frac{\text{var } d}{\text{var } y}$$

where var is the variance which is the sum of squared deviations from the average divided by the number of observations

There are other measures used in MRA to express confidence intervals for expected values and actuals, and the F value for the analysis of variance.

If more than two variables are present, the mathematics can become very involved, and therefore computerized techniques should be used. Most statistical packages on microcomputers will have an MRA feature.

When undertaking MRA, there may be correlation between the values, but they may not in fact be related; for instance, there was once a study done comparing the number of bananas imported into Sweden and the number of live births with good correlation!

It is also important to use the services of a trained statistician to check the feasibility of the results and to explain what the various measures portray.

In practice the technique is used to calculate, for instance, the total time needed to perform the tasks when there are A invoices, O orders and C cheques. Then perhaps with time T would be given in the equation

$$T = 1.87 + 0.56A + 0.270 + 0.15c$$

which can be calculated.

The regression equation would be solved needing a large number of observations in the office before specific variables could be allocated to the above formulae.

10.8 Work standards for maintenance

The basic techniques for measuring indirect work, which includes maintenance, are similar to those used for measuring direct work. These are usually broken down into:

1. Analytic estimating.
2. Time study.
3. Predetermined time standards.

There are many companies which have made significant reductions in maintenance costs through installing work standards linked to method improvement studies.

Work measurement is not a substitute for good supervision. If work is not planned and scheduled then labour effectiveness will suffer. Work measurement does not guarantee work of high quality, but quality should not suffer because of work measurement if supervisors are doing their jobs properly.

UNIVERSAL MAINTENANCE STANDARDS (UMS)

Setting accurate maintenance standards with time study is too costly to be practical, requiring as it does almost a one-to-two ratio of time study men to maintenance men. Standard data or time formulas based on the study offer a better solution. This approach has been known for years but has not until recently been used extensively because the cost of collecting all the data together in the first place is very high indeed.

Time standards usually provide a standard time for doing the job. However, it is not practical to expect all maintenance workers to do a given job with exactly the same motion patterns in exactly the same time. In practice, most maintenance men will not accept as a fact that a standard for replacing a valve in a pipeline is, for example, exactly 28.5 minutes. They will agree that a job can be performed in, say, between 20 and 40 minutes. It is on this concept that UMS is based and its introduction is not too difficult if the following approach is used:

1. Develop accurate time formulas for every type of work performed by the maintenance department.
2. Establish standard work groupings.
3. Establish benchmark jobs.

STANDARD GROUPINGS

Studies of maintenance work almost invariably show that about 80 per cent of the maintenance jobs require less than eight hours to perform. The numerous short jobs cause the real standard-setting problem. One answer to this is to establish standard work groupings; an example is given in Table

Table 10.2 Example of standard work grouping

Group	Range, min	Mean, min
A	0.00 to .015	0.10
B	0.15 to 0.25	0.20
C	0.25 to 0.50	0.40
D	0.50 to 0.90	0.70
E	0.90 to 1.50	1.20
F	1.50 to 2.20	2.20
G	2.50 to 3.50	3.00
H	3.50 to 4.50	4.00
I	4.50 to 5.50	5.00
J	5.50 to 6.50	6.00
K	6.50 to 8.00	7.30
L	8.00 to 10.00	9.00

10.2. The job of the analyst is to take a close look at each job and place it in the particular group to which it belongs.

BENCHMARK JOBS

The allocations of different, but typical, maintenance jobs carried out in a specific plant are then slotted into the particular groups. This is so that standards of comparison can be made by the analyst when setting up a UMS system.

COMPILING UMS STANDARDS

There are three other considerations to be applied in the time per job:

1. *Job preparation time* For example, receiving instructions, gathering tools and equipment, preparation at job site, collecting and cleaning job site.
2. *Travel time* The standard set for travelling to different time zones.
3. *Allowances* Include delays of any type: delay in activating personnel, unavoidable delays, balancing delays, planning time, etc.

Experienced personnel are necessary in setting UMS standards, and should not be permitted to practise until formal UMS training schemes have been undertaken, otherwise the scheme could run into disrepute very quickly indeed.

PREVENTATIVE MAINTENANCE (PM)

Ideally, all maintenance should be preventative. That is, all maintenance should be performed prior to any equipment failure. In these terms 'failure' means the point at which there is a deterioration in quality or quantity of the product. At the outset of any PM programme, however, it is quickly realized that a situation where all maintenance is preventative is not economically feasible, since equipment would be grossly over-maintained.

Different applications of PM will be made in different plants, depending upon whether output can be put into storage; if it can then PM is at a lower level than when output has to be fed directly to customers.

Some managements view PM as merely lubrication, painting and cleaning operations, normally conducted on a scheduled basis. Others expect PM programmes not only to prevent downtime but also to minimize costs, improve output, and influence the quality of the product.

Since PM actually starts before the equipment is built or purchased, maintenance considerations are applied from the very beginning, including the diagnosing of the faults that cause equipment to fail; this forms the basis of the PM programme.

10.9 Work measurements linked to incentives

Money is not the only motivator at work. However, just about everybody works for money—very few would actually come to work if they received little or no money. Money is an excellent motivator; other factors are also motivators or demotivators.

It is generally the experience of industrial engineering practitioners that if people are working without incentives then their work rate can be increased by 50 to 100 per cent without any detrimental effect on quality.

TYPES OF INCENTIVE SCHEMES

These are:

1. Piecework.
2. Premium: (a) straight line;
    (b) variations on (a), e.g., $50 + 0.5$;
    (c) curve-based schemes.
3. Time saved, e.g., $\dfrac{\text{allowed time} - \text{time taken}}{\text{time taken}}$.
4. Measured daywork—performance bands.
5. Multi-factor schemes, e.g., performance, quality, timekeeping and delivery indexes.
6. Value added schemes based on value added ÷ total wages or value added ÷ total attendance time.

INDUSTRIAL ENGINEERING

7. Profit-sharing schemes.
8. Total productivity schemes, e.g., Improshare.

PROBLEMS OF COST

The main problem in the past has been the costs of administration for incentive schemes, especially:

1. Cost of measurement.
2. Cost of calculating and controlling the incentive.

The cost of measurement has been drastically reduced through modern techniques, e.g., MOST, computerized systems and synthetics. The cost of administration can also be reduced sharply by using microcomputer systems of shop floor data collection and/or monitoring or by using programs for calculating incentives. Such systems are now becoming cost-effective.

IMPROSHARE

This particular incentive scheme was devised by Mitchell Fein, an American industrial engineer, in 1974. The name Improshare means 'improved productivity through sharing'. It has been implemented in many hundreds of different companies both in the United States and in Europe. In the United States it has yielded an average increase of 35 per cent in shop floor productivity while some applications have achieved over 100 per cent. In the United Kingdom the consultancy company Arthur Young, McClelland and More has the rights for installing this proprietary system.

The basic premises

The basic features of Improshare are as follows:

1. Benefits derived from productivity increases are shared on an equal 50:50 basis between the employees and the company.
2. Bonus is earned as soon as productivity increases above the existing level. The exact productivity measure will depend upon the needs of the company, but generally it is measured by the output hours of work produced divided by the hours of input for a given period.
3. The methods used by the company to evaluate its own output can remain unchanged, so long as they do provide a consistent and reliable means of measurement.
4. Productivity teams are formed of voluntary groups (similar to quality circle groups) who meet regularly to identify and analyse problems occurring in their own work area.

5. A limit on the agreed maximum weekly bonus payment is agreed; any earnings above this are banked, to be drawn out on lower productivity weeks. Where this is constantly exceeded a lump sum is paid out and the base level of the plan is adjusted.
6. Agreements are made on how to share out productivity changes resulting from capital investment projects, the bonus paid out normally being a percentage of the savings arising from investment.

10.10 Conclusion

Work measurement is a foundation technique for modern operations management. Besides helping to achieve higher productivity of the resources of manpower, capital and plant, it provides basic data for ascertainment and control of costs through modern techniques of management accounting. It also enables balanced production programmes to be applied to achieve optimum utilization of labour and machines, subject to the limitations imposed by availability of materials and the order book position.

PRODUCTIVITY THE KEY

'Without productivity objectives a business does not have direction. Without productivity measurement, it does not have control.'—Peter Drucker.

Part 4

Introduction to statistical method and operations research

11
Introduction to statistical method

11.1 Definitions

1. *Statistical data*: numerical information.
2. *Statistics*: a body of methods for making wise decisions in the face of uncertainty.
3. Statistical techniques are based on ideas about the nature of variation; we use numerical information to operate these ideas.
4. Statistical data are used for two purposes:

    (a) Practical action.
    (b) Scientific knowledge.

5. The essential difference between these two types of purpose is that in the first we can list a limited number of plans of action which can be considered; consequences of error here can be evaluated. In the second the purposes are unspecified and so consequences of error cannot be readily evaluated.
6. Statistical processes may be summarized as follows:

    (a) Collection of data.
    (b) Tabulation of data.
    (c) Interpretation of data.

7. Complete information, generally speaking, is not available. Statistics provides rational principles and techniques which tell when and how judgements may be made on the basis of partial information, and what partial information is worth seeking.

11.2 Statistics and scientific method

There are four stages in problem-solving by scientific method, as follows:

1. *Observation*   Facts relevant to the problem are collected and studied.

2. *Hypothesis* To explain the facts observed, a hypothesis or theory is formed expressing the patterns which may have been detected in the data.
3. *Prediction* Deductions are made from the theory; these constitute new knowledge if the theory is satisfactory. If the theory is to be of value then it must make possible such new knowledge. The theory should make it possible to anticipate what will be seen if certain observations, not yet made, are made.
4. *Verification* New facts are collected to test the predictions made. With this step the cycle starts all over again.

There is no final truth in science because, although failure to refute a theory increases confidence in it, no amount of testing can absolutely 'prove' that it will always hold.

Statistics are helpful in the first stage as a guide to the most useful information to be obtained and to how the results may be interpreted.

1. The degree of confidence in the conclusion and the necessary allowance for error are emphasized by the statistician.
2. The branch of statistics which helps to summarize and classify data is called 'descriptive' statistics.
3. 'Analytical' statistics deals with methods of planning the observation and analysing and basing decisions on the data so summarized.
4. Statistical techniques use numerical data. These can be obtained even when qualitative or subjective material is the subject of study, e.g., quality.
5. In many situations it is possible to predict patterns of events as opposed to individual events. This is true even in natural science where many theories are stochastic rather than deterministic in nature. It is particularly true in the realm of atomic phenomena such as radioactivity and in quantum mechanics.
6. The essential question at the verification stage is, 'Is the discrepancy reasonably attributable to change?' If it is then there are no reasonable grounds for looking for a 'special' cause. In quality control this is called an 'attributable' cause. This is the kind of judgement we are called upon to make when carrying out tests of significance and when we use a control chart.

12

Descriptive statistics

12.1 Introduction

It is helpful at the outset to define just what we mean by statistics: Statistics is concerned with scientific methods for *collecting, organizing, summarizing, presenting,* and *analysing* data, as well as *drawing valid conclusions and making reasonable decisions* on the basis of such analysis.

These areas are now explored in more detail.

12.2 Collecting data

Fundamental to all data collection is the process of sampling: Sampling is the process of drawing inferences concerning the characteristics of a mass of items by examining closely the characteristics of a somewhat smaller number of items drawn from the entire mass.

TYPES OF SAMPLING

There are basically three types of sampling in common use:

1. Random sampling.
2. Systematic sampling.
3. Selective or stratified sampling.

We will be concerned only with the first two of these:

Random sampling

In the technique of activity sampling, random sampling is used as the basis for collection of data. A random sample is one in which every member of the population* has the same chance of being included in the sample. (In activity

---

*'Population' is a term used in statistics to describe the entire field or area from which the sample can be drawn.

sampling, random-number tables or lottery methods are used to ensure that this condition is complied with.)

Systematic sampling

A systematic sample is one in which every ninth number of the population is selected. The starting point is chosen at random, but thereafter the constant interval n determines the rest of the sample. For example:

1. If every tenth file were withdrawn from a filing cabinet to check the contents then this would be a systematic sample.
2. In the technique of constant internal activity sampling, a record of the actions of the operators is taken, say, every 30 seconds. This is in effect a systematic sampling process.

SAMPLING ERROR

As a result of using a sampling process the answer obtained will differ from the true value by some amount of error. This can be of two types:

1. Sampling error.
2. Non-sampling error.

Sampling error

This arises because only part of the population is observed. It can be kept to a minimum by suitable survey design and its size estimated.

Non-sampling error

This type of error arises in observing populations as well as samples and may arise from any combination of the following:

1. Faulty questionnaire design.
2. Incompetence of investigators.
3. Incorrect answers by respondents.
4. Sample not representative of intended population.
5. Errors in analysis of data.

This type of error is impossible to estimate and control. The survey should be so designed that the total error is minimized.

DESCRIPTIVE STATISTICS

ADVANTAGE OF METHODS OF SAMPLING AS OPPOSED TO OBSERVATION OF TOTAL POPULATION

1. Reduced costs—only part of the field observed.
2. Greater accuracy—possibility of using smaller number of trained investigators.
3. Greater scope—possibility of obtaining quite elaborate information by concentrating on each individual or situation.
4. Great speed—administration easier and results obtained more quickly.

PRINCIPAL STEPS IN A SAMPLE SURVEY

1. Statement of:
   (a) What is to be achieved.
   (b) Population to be covered.
   (c) Information to be collected.
   (d) Precision required.

2. Methods of collecting information:
   (a) Personal interview.
   (b) Postal questionnaire.

3. Selection of samples: type and size.
4. Pilot or preliminary survey.
5. Obtaining actual data.
6. Dealing with data collected.

12.3 Frequency and cumulative frequency distributions

A frequency distribution (f.d.) of a group of observations is formed by setting up some classification scheme and counting how many observations fall into each class. The classification scheme usually satisfies the restriction that every observation can fall into one class, and only one. However, this restriction could be broken depending on the use to which the classification is to be put.

If the observations are numerical measurements, *and* the classification is by size (of the measurement), *and* the restriction referred to above has been adhered to, then a cumulative frequency distribution can be derived from a frequency distribution; this is explained in detail later in this chapter.

FREQUENCY AND PERCENTAGE FREQUENCY TABLES

The figures in Table 12.1 are measures of the efficiency of a group of men over a period of 40 consecutive weeks. If we want to answer such questions as, 'Do

Table 12.1 Efficiencies of a group over 40 weeks

80	83	71	83	78	77	74	84	75	78
72	88	84	80	73	82	81	73	83	81
77	84	79	81	89	74	85	80	82	76
84	80	76	87	76	79	80	86	79	78

the men ever work at an efficiency above 80 for more than three consecutive weeks?' or, 'Is a "good" week usually followed by a "bad" week?', then the arrangement of these figures as in Table 12.1 is quite useful. On the other hand, if we want to answer such questions as, 'How often do they work at below 80 per cent efficiency? or, 'What is the average [see page 105] efficiency they work at?' then the arrangements shown in Table 12.2 or 12.3 are more useful.

The basic steps in the construction of a frequency table from a group of measurements are:

1. Find the smallest and largest member of the group.
2. Find the total number of measurements.
3. Decide on a classification scheme. The class could be *either* of the following:

   (a) A list of all the different values of measurements which occur, in which case we would end up with an 'ungrouped' frequency table.

   (b) A series of intervals which cover the whole range (from the smallest to the largest) of measurements, in which case we would end up with a 'grouped' frequency table. An interval is represented by two numbers; the smaller is called the lower limit, and the larger the upper limit. (It will be seen that an ungrouped frequency table is just a special case of a grouped frequency table, which has each interval with the lower limit equal to the upper limit.) Unless there is a special case for not doing so, the difference between the two limits should be the same for each interval. Just how many intervals there should be depends on the use the frequency table will be put to, and (sometimes) the number of observations to be classified; however, the following considerations should be noted:

   (i) Few intervals imply a too general picture of the distribution with too much information being lost.

   (ii) Many intervals might give too much detail, making it impossible

DESCRIPTIVE STATISTICS

Table 12.2 Frequency table derived from Table 12.1

Efficiency range	Frequency
70–73	4
74–77	8
78–81	14
82–85	10
86–89	4

Table 12.3 Percentage frequency table derived from Table 12.2

Efficiency range	Percentage frequency
70–73	10
74–77	20
78–81	35
82–85	25
86–89	10

to get the general features of the distribution without further calculation.

Generally speaking, the number of intervals should be between 4 and 20, and the average number of observations per interval should not be less than 4. One last consideration may save a lot of work: if the numbers involved are awkward to handle, then try to choose the lower limit of the first interval and the width of the interval (the difference between the lower limit of one interval and the lower limit of the next), so that once the first interval has been defined the other intervals are calculated easily.

4. Count how many observations fall into each class. Collecting the results together, and presenting them in a table with the intervals in ascending order, gives a frequency table (Table 12.2).

A percentage frequency table is calculated from a frequency table by dividing each figure in the frequency column by the total frequency and multiplying by 100 (Table 12.3).

CUMULATIVE FREQUENCY AND CUMULATIVE PERCENTAGE FREQUENCY TABLES

When a measurement is written down this is often the approximation of some other, more accurate, number. The most common methods of approximation are:

1. 'Rounding to the nearest (integer, tenth, etc.—depending on the level of accuracy required).' If this method had been used when compiling Table 12.1 then a figure of, say, 84, would represent an approximation of anything from 83.5 to 84.49.
2. 'Rounding down to the nearest (integer, etc.).' Had this method been used

INDUSTRIAL ENGINEERING

Table 12.4 Cumulative frequency table formed from Table 12.2

Efficiency	Cumulative frequency
73.5	4
77.5	12
81.5	26
85.5	36
89.5	40

Table 12.5 Cumulative percentage frequency table formed from Table 12.4

Efficiency range	Cumulative percentage frequency
73.5	10
77.5	30
81.5	65
85.5	90
89.5	100

in compiling Table 12.1 then 84 would represent an approximation of anything from 84 to 84.9.

Bearing these remarks in mind, it is useful to define the following terms in connection with intervals in a frequency table.

The *least upper bound* (l.u.b.) of an interval is the lowest number which is greater than or equal to any of the possible (accurate) values which could have their approximated value included in that interval.

Similarly, the *greatest lower bound* (g.l.b.) of an interval is the greatest number which is less than or equal to any of the possible (accurate) values which could have their approximated value included in that interval. The 'width' of an interval is the difference between the l.u.b. and the g.l.b.

A cumulative frequency table is formed from a frequency table by calculating the l.u.b. of each interval, counting how many measurements are less than or equal to it, and arranging this information in a table in ascending order of the l.u.b.s. (It will be seen that the method of approximation assumed for the calculation of the l.u.b.s for Table 12.4 is method 1.)

A cumulative percentage frequency table is formed from a cumulative frequency table by dividing each item in the cumulative frequency column by the total frequency and then multiplying by 100 (Table 12.5).

HISTOGRAMS

A histogram can be constructed from a frequency table (or percentage frequency table) only when the l.u.b. of each interval is equal to the g.l.b. of the succeeding one. The steps in the construction of a histogram (Fig. 12.1) are:

1. Draw a horizontal line intersecting a vertical line.
2. Decide on a suitable scale for frequency on the vertical line.

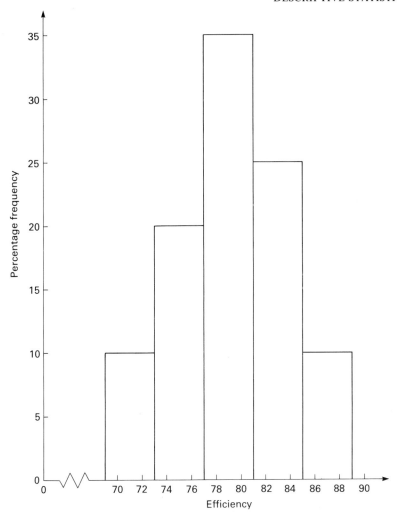

Figure 12.1   Histogram formed from Table 12.3

3. Decide on a suitable scale for the horizontal line to cover the full range of measurements, from the g.l.b. of the first interval to the l.u.b. of the last.

   Note: If either scale is not continuous (as in Fig. 12.1) then it is conventional to break the line in some way to indicate this, since a discontinuous scale (especially frequency) can give an extremely misleading picture.

4. Using the g.l.b. and l.u.b. as a 'base', construct a 'bar' for each interval. If

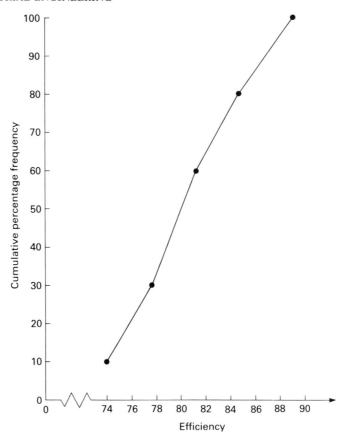

Figure 12.2 Ogive constructed from Table 12.4

the intervals are all of equal width then the height of a bar should be equal to the frequency value for that interval. If the intervals are of unequal widths, some (any) fixed width of interval is chosen and the height of the bar for each interval is calculated by the following formula:

$$\text{Height} = \frac{\text{fixed width}}{\text{interval width}} \times \text{frequency}$$

OGIVES

An ogive is a graph of the cumulative or cumulative percentage frequency distribution. The most useful of these is the graph of the cumulative percentage f.d. (Fig. 12.2). It can be drawn directly from the table of cumulative percentage f.d. (Table 12.5).

DESCRIPTIVE STATISTICS

12.4 Parametric representation

A 'parameter' is a number which measures some characteristic of a group of numbers.

MEASURES OF LOCATION

There are two kinds:

1. Measures of central tendency.
2. Fractiles.

Measures of central tendency or averages

The word 'average' is used in the English language to describe a number that is in some way more representative of a group of numbers than any other. A shipbuilder might say the 'average' size of ship he builds is 100 000 tonnes; a salesman might say the 'average' value of his monthly sales is £20 000; a psychologist might say the 'average' score for his IQ test is 100; yet they would all probably use different methods to calculate this average figure. There are three common methods of calculating an average, and the parameters arrived at by these three methods are called the arithmetic mean, the median, and the mode.

Arithmetic mean (a.m.)

The a.m. of a group of measurements is obtained by summing the measurements and dividing the sum by the total number of measurements in the group. If x represents any measurement in the group, then the a.m. is represented by $\bar{x}$, and is given by the formula

$$\bar{x} = \frac{\sum x}{n}$$

where n is the number of measurements in the group
An equivalent, and often more useful, formula is

$$\bar{x} = \frac{\sum (x-c)}{n} + c$$

where c is any constant

Median

The median is that measurement below which half the measurements lie. If n (number of measurements in the group) is odd then the median is the middle measurement when they are listed in order of size. If n is even then the median is the a.m. of the two middle measurements. (See section 12.5 for details of the calculation of the median from an ogive.)

Mode

The mode of a group of measurements is that one which is repeated most often. If the number of different measurements is large then it is usually not useful to use this definition, and the mode is defined with respect to a histogram (see section 12.5 for details). The 'modal class' of a frequency distribution is that class which occurs most frequently.

Fractiles

Four different types of fractiles are distinguished:

Median

See above.

Quartiles

These divide the group into quarters. The 'first' or 'lower' quartile is that value below which lies the bottom 25 per cent of the group.

The second quartile is the median.

The third quartile is that measurement below which lies the bottom 75 per cent of the group.

Deciles

These divide the distribution into tenths.

The first decile is that value below which lies the bottom 10 per cent of the group.

The second decile is that value below which lies the bottom 20 per cent of the group... and so on up to the ninth decile.

Percentiles

These divide the distribution into hundredths, e.g., the 87th percentile of that value below which lies the bottom 87 per cent of the group. They are used only when the number of observations is very large, e.g., government survey data, national exam results or psychological test scores.

Fractiles can be calculated very easily from the ogive (see section 12.5). They

DESCRIPTIVE STATISTICS

can also be obtained by linear interpolation from a cumulative frequency distribution.

MEASURES OF DISPERSION

Most measures of dispersion are concerned with indicating the spread of values about the a.m. of the group.

Mean average deviation (m.a.d.)

This is the a.m. of the difference (made positive) between each measurement and the a.m. of the group. If x represents a typical measurement then the m.a.d. is given by the formula

$$\text{m.a.d.} = \frac{\sum |x - \bar{x}|}{n}$$

The upright bars mean 'forget the minus sign when the result of the calculation between the bars is negative'. See section 12.5 for details of the calculation of the m.a.d. from a frequency table.

Interquartile range and semi-interquartile range

The interquartile range is equal to the difference between the third and first quartiles.

Range

This is the difference between the largest and smallest member of the group. For large groups it is likely to give a distorted picture because of extreme values; however, because of its ease of calculation, it is sometimes used on small samples (up to 12 items) for constructing quality control charts.

Variance and standard deviation (s.d.)

These are related by the formula, variance $= (\text{s.d.})^2$. Almost all practical statistical techniques have a theoretical justification. Because these two measures are easiest to work with in theoretical statistical work, nearly all prescriptions for the application of statistical techniques which need a measure of dispersion use these. The first three measures are normally used for descriptive work, these are used for analytical work.

The variance of a group of measurements is the a.m. of the square of the differences between each measurement and the a.m. of the group. If x

INDUSTRIAL ENGINEERING

represents any measurement, then the variance (usually denoted by σ^2 while the s.d. is denoted by σ) is given by the formula:

$$\sigma^2 = \frac{\sum(x-\bar{x})^2}{n}$$

where n is the total number of measurements.

An equivalent, and often more useful, formula is

$$\sigma^2 = \frac{\sum(x-c)}{n} - (\bar{x}-c)^2$$

where c is any constant (note that if c is zero, this formula reduces to

$$\sigma^2 = \frac{x}{n} - \bar{x}^2$$

and if c is 0, this formula is the same as the first one).

Coefficient of variation (c.v.)

This is given by the formula

$$\text{c.v.} = \frac{\text{s.d.}}{\text{a.m.}} \times 100$$

This is always used to compare variation in two or more groups of measurements where the variance is expected to increase with larger values of measurement, e.g., a firm might want to compare the variability of production between its North East factory, which produces an average (a.m.) of, say £20 000 of good per week, with its South East factory, which produces an average of, say, £30 000 of goods per week.

12.5 Appendix

CALCULATION OF A MODE FROM A HISTOGRAM

1. Find the highest bar. (If there are two bars which are highest, the distribution is bi-modal; if there are three or more then the distribution is said to be multi-modal.)
2. Draw a line from the top left corner of the highest bar to the top left corner of the adjacent bar to its right. Exceptions:

    (a) If there is no adjacent bar, draw a diagonal, i.e. a line from the top left corner to the bottom right.
    (b) If the histogram has two bars of equal height and these are adjacent, treat these two bars as one wider bar.

DESCRIPTIVE STATISTICS

3. Similarly, draw a line from the top right corner.
4. Draw a vertical line from the point of intersection of these two lines: the point where it intersects the horizontal axis is the mode of the distribution.

Figure 12.3 shows these steps carried out on the histogram in Fig. 12.1.

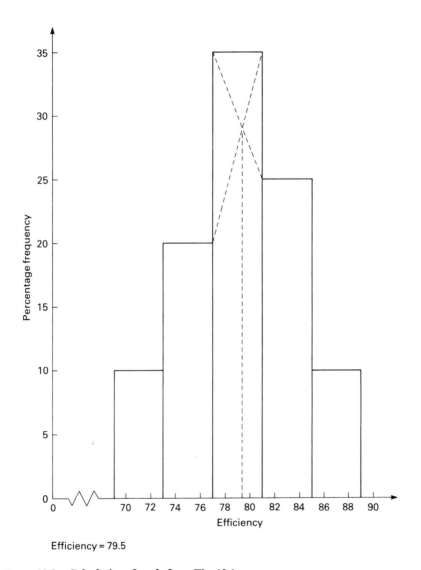

Efficiency = 79.5

Figure 12.3 Calculation of mode from Fig. 12.1

INDUSTRIAL ENGINEERING

CALCULATION OF A FRACTILE FROM AN OGIVE

1. Note from the definition of the fractile you wish to calculate what percentage of a distribution is below this fractile.
2. Find this percentage on the vertical axis of your graph and draw a horizontal line through it to meet the ogive.
3. Where this horizontal line meets the ogive, draw a vertical line to meet the horizontal axis: where they meet gives the value of the fractile you wish to calculate.

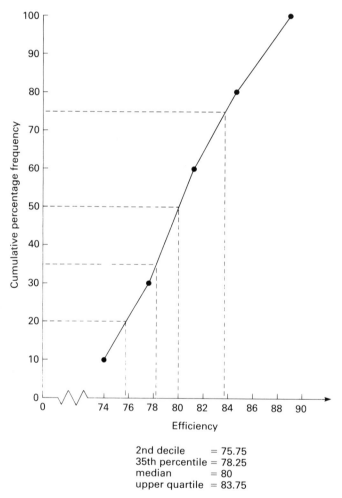

2nd decile   = 75.75
35th percentile = 78.25
median   = 80
upper quartile = 83.75

Figure 12.4 Calculation of fractile from Fig. 12.2

DESCRIPTIVE STATISTICS

Figure 12.4 shows these stops carried out on the ogive shown in Fig. 12.2 to calculate the second decile, 35th percentile, median and upper quartile.

CALCULATION OF A.M., M.A.D., VARIANCE AND S.D. FROM A FREQUENCY TABLE

For each of these calculations, the following definition is required:

The *mid-point* of an interval is the a.m. of the g.l.b. and the l.u.b. of the interval.

The following calculations are all done on Tables 12.6 to 12.8. In the formulas, these abbreviations are used:

- f denotes the frequency of an interval.
- x denotes the mid-point of an interval.
- $\bar{x}$ denotes the a.m. of the distribution.
- c denotes any constant.

Calculation of the a.m. using the formula

$$\bar{x} = \frac{\sum f(x-c)}{\sum f} + c$$

See Table 12.6.
In the table, since it is the figures in the '$x-c$' column which are used in the formula, c has been chosen as 71.5 to make the number more manageable.
From Table 12.6 we see that:

$$\sum f = 40, \quad \sum f(x-c) = 328, \quad c = 71.5$$

Substituting these values in the formula gives

$$\bar{x} = \frac{328}{40} + 71.5 = 8.2 + 71.5 = 79.7$$

Table 12.6 Table for calculation of the arithmetic mean

Interval	f	x	x−c (c = 71.5)	f(x−c)
70–73	4	71.5	0	0
74–77	8	75.5	4	32
78–81	14	79.5	8	112
82–85	10	83.5	12	120
86–89	4	87.5	16	64
Σ	40		Σ	328

INDUSTRIAL ENGINEERING

Table 12.7 Table for calculation of the mean average deviation

Interval	f	x	$x - \bar{x}$	$f|x - \bar{x}|$
70–73	4	71.5	8.2	32.8
74–77	8	75.5	4.2	33.6
78–81	14	79.5	0.2	2.8
82–85	10	83.5	3.8	38.0
86–89	4	87.5	7.8	31.2
Σ	40		Σ	138.4

Calculation of the m.a.d. using the formula

$$\text{m.a.d.} = \frac{\Sigma f|x - \bar{x}|}{\Sigma f}$$

See Table 12.7. The same group of numbers is used here as on page 111.
From Table 12.7, $\Sigma f = 40$, $\Sigma f|x - \bar{x}| = 138.4$; substituting these values in the formula gives

$$\text{m.a.d.} = \frac{138.4}{40} = 3.46$$

Calculation of the variance (and s.d.) using the formula

$$\text{variance} = \frac{\Sigma f(x - c)^2}{\Sigma f} - (\bar{x} - c)^2$$

See Table 12.8. In the table c has been chosen as 79.5: this makes the calculations even easier than choosing $c = 71.5$ as was done on page 111.

Table 12.8 Table for calculation of the variance and standard deviation

Interval	f	x	$x - c$ ($c = 79.5$)	$(\bar{x} - c)^2$	$f(x - c)^2$
70–73	4	71.5	−8	64	256
74–77	8	75.5	−4	16	128
78–81	14	79.5	0	0	0
82–85	10	83.5	4	16	160
86–89	4	87.5	8	64	256
Σ	40			Σ	800

From the table, $\sum f = 40$, $\sum f(x-c)^2 = 800$; from previous pages $\bar{x} = 79.7$, so $(\bar{x}-c)^2 = (79.7-79.5)^2 = 0.2^2 = 0.04$.

Substituting these values in the formula gives

$$\text{variance} = \frac{800}{40} - 0.04 = 20 - 0.04 = 19.96$$

Since s.d. $= \sqrt{\text{variance}}$, then s.d. $= \sqrt{19.96} = 4.47$.

13

The normal distribution

13.1 Introduction

This is almost certainly the most useful probability distribution in the entire field of statistics. It was first discovered by the English mathematician De Moivre (1667–1754), and later rediscovered and applied in science (both physical and social) and in practical affairs by the French mathematician Laplace (1749–1829). It was also extensively developed and utilized by the German mathematician Gauss (1777–1855). This distribution is particularly important because measures computed from samples usually tend to follow a normal distribution whether or not the original data are normally distributed.

13.2 The importance of the normal distribution

The normal distribution is important in its own right. If a large number of careful measurements are made of a single magnitude, say the length of a steel bar or the resistance of an electrical circuit, these measures will strongly tend towards a normal distribution. The values found will concentrate around the mean and thin out toward both the higher and lower values. Hence the normal distribution is central to the theory of measurement and consequently basic to all careful scientific work.

A very large number of natural distributions seem to be close approximations to the normal. The distribution of almost any dimension produced by almost any machine tool tends to be normal. The heights and weights of people are approximately normally distributed. It is also believed that human intelligence follows the normal distribution approximately.

The normal distribution is by far the most useful distribution in the field of 'inferential statistics'. This field of statistical analysis, in which inferences about certain characteristics of a 'population' are drawn from measurements made on small samples, very much relies on the properties and features of the normal distribution.

The normal distribution also serves as a good approximation to a number of other theoretical distributions, which have probabilities either too

THE NORMAL DISTRIBUTION

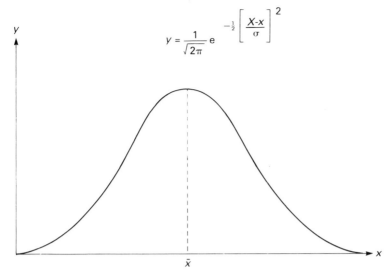

Figure 13.1 Normal distribution curve

laborious or impossible to work out exactly. Two good examples in this respect are the binomial and Poisson distributions, which under certain conditions can be approximated by the normal distribution.

13.3 Basic features of the normal distribution

Any normal distribution curve (Fig. 13.1) has the following features:

1. It is symmetrical.
2. It is bell-shaped.
3. Its mean lies at the peak of the curve.
4. The two tails never actually touch the horizontal axis, although they continuously approach it.
5. The formula for the curve is:

$$y = \frac{1}{\sqrt{2\pi}} e^{-\frac{1}{2}\left(\frac{X-\bar{x}}{\sigma}\right)^2}$$

(Students may ignore it completely at this stage, as any data relating to the curve can be easily obtained from statistical tables.)

6. Any normal distribution is fully defined by two parameters, $\bar{x}$ the mean and σ the standard deviation. The arithmetic mean determines the location of the curve (see Fig. 13.2) and the standard deviation determines the shape of the curve (see Fig. 13.3).

115

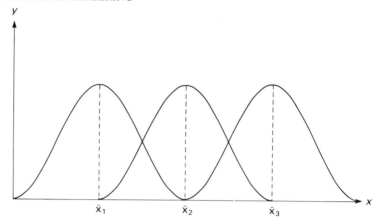

Figure 13.2 Arithmetic mean and location of normal distribution curve

13.4 Areas below the normal distribution curve

The most important aspect in connection with using the normal distribution curve is in determining the area under the curve. This area represents the probability of occurrence of a variable selected at random from the normal distribution.

There are broadly two approaches to measuring this area. The first is by computation. Mathematicians, in fact, compute the area that lies between any two lines, the curve and the axis, by using the formula for the curve, mentioned earlier. This method is, however, extremely laborious and does need a fair amount of mathematical knowledge in operating it. The second method, which is extremely useful for most practical purposes, is through the

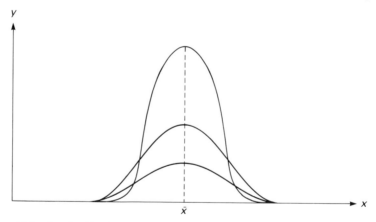

Figure 13.3 Standardization and shape of normal distribution curve

THE NORMAL DISTRIBUTION

use of statistical tables. In this method, the calculation of the area becomes merely a matter of determining the appropriate values from tabulated data.

There is a fairly wide variation in the way in which the areas under the normal distribution curve have been tabulated by various authors. The following remarks refer to *Statistical Tables* by Murdoch and Barnes (Macmillan, 2nd edn, 1971).

In the calculation of all areas under the normal distribution curve the following must be remembered:

1. The total area under the curve equals unity.
2. The area under the curve on either side of the mean value is equal to 0.5000.
3. For calculation purposes the deviations measured on one side of the mean are denoted as positive and those on the other as negative. However, when areas on both sides of the mean are being considered in the same analysis then the signs are ignored and the areas are added together.
4. All statistical tables have values of areas tabulated for what is called a 'standardized normal variable', whose mean value is zero and whose deviation from the mean is measured in terms of the standard deviation.

With reference to the tables by Murdoch and Barnes (see Fig. 13.4), values of the shaded area under the curve are tabulated for different values of the deviation z. For instance, for a value of $z = 1.96$ the corresponding area tabulated is 0.025. This can be interpreted as follows:

1. The area under the curve beyond the point 1.96 deviations from the mean is 0.025.

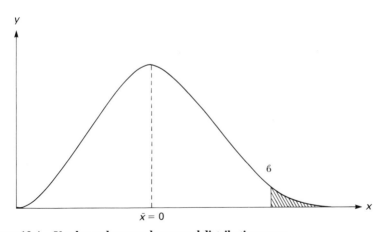

Figure 13.4 *U*-value and area under normal distribution curve

2. The area under the curve between the mean and the point 1.96 deviations from the mean is $0.500 - 0.025 = 0.475$.

Thus for most practical purposes such a table will provide all the values of areas required for a wide range of conditions.

Before using such tables for practical applications, a small conversion is necessary. This is because in practice one can come across various normal distributions with different values of means and standard deviations. All of these values will have to be converted to the scale used in the table, namely that of the mean being zero and the standard deviation being unity. This conversion process is done by using the following relationship:

$$z = \frac{x - \bar{x}}{\sigma}$$

where $\bar{x}$ and σ are the mean and standard deviation values for the actual distribution and x is the deviation from the mean in the actual distribution.

Finally, it must be remembered that since the area under the curve represents probabilities, they can be expressed as percentages. For example, in the case quoted above, $z = 1.96$, area $= 0.025$, the following interpretation can be given:

1. The probability of a random variable lying beyond the point of $+1.96$ deviations from the mean is 2.5 per cent.
2. The probability of a random variable lying between the mean and the point of $+1.96$ deviations from the mean is 47.5 per cent.

13.5 Important features of the normal distribution curve

The following features of the normal distribution are worth noting (see Fig. 13.5):

1. The area under the normal curve between the mean $+1$ standard deviation and the mean -1 standard deviation is 68 per cent of the total area.
2. The area under the normal curve between the mean $+2$ standard deviations and the mean -2 standard deviations is 95 per cent of the total area.
3. The area under the normal curve between the mean $+3$ standard deviations and the mean -3 standard deviations is 99.7 per cent of the total area.

These are approximations, but are sufficient for most work in many practical managerial situations.

THE NORMAL DISTRIBUTION

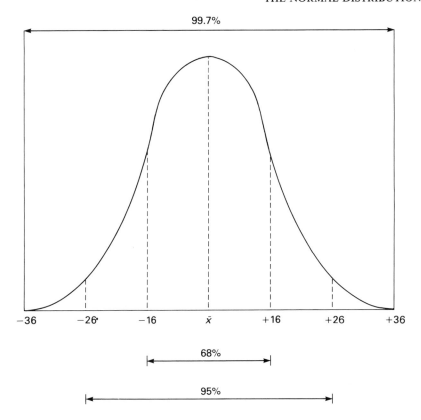

Figure 13.5 Important features of normal distribution curve

The relationship between probability and the area under the normal curve enables one to make a statement and know just how often it will prove to be true.

For instance, suppose that 30 items were selected *at random* from a large normal distribution. How many of the 30 do you think would have values that lay within 1 standard deviation of the mean? We know that approximately 66.667 per cent of the total area lies below a normal curve within 1 of the mean. This means that 2 out of 3 items lie within 1 of the mean, and therefore 20 out of our 30 will probably have values within 1 of the mean. Similarly, since 95 per cent lie within 2 of the mean, then 19 out of 20 (95 per cent) of the selected items will be within 2 of the mean. Hence if we selected just one item at random we could assert it was within 2 of the mean, knowing that if we were to test our assertion often enough it would be correct 'at a 95 per cent confidence level'.

INDUSTRIAL ENGINEERING

EXAMPLE

A process produces components with a mean of 2 inches and a standard deviation of 0.001 inch.

1. What proportion of components will be larger than 2.002 inches?
2. What proportion of components will be between 1.998 and 2.001 inches?
3. What proportion of components will be outside the tolerance limits of 2.000 ± 0.0025 inches?

Solution

1.
$$U = \frac{2.002 - 2.000}{0.001} = +2.0$$

Reference to tables gives the area beyond $U = 2.0$ as 0.02275.
 Therefore proportion of components larger than 2.002 inches is 2.275 per cent.

2. This has to be worked in two parts:
   (a) Proportion larger than 2.001 inches:
   $$U = \frac{2.001 - 2.000}{0.001} = 1$$

   Therefore proportion larger than 2.001 = 0.1587.

   Therefore proportion between
   2.000 and 2.001 = 0.5 − 0.1587 = 0.3413.
   (b) Proportion smaller than 1.998 inches:
   $$U = \frac{1.998 - 2.000}{0.001} = -2.0$$

   From the symmetry of the normal curve about its mean,

   Proportion smaller than 1.998 = 0.02275.

   Therefore proportion between
   1.998 and 2.000 = 0.5 − 0.02275 = 0.47725.
   Therefore proportion between 1.998 and 2.001 = 0.81855 or 81.9 per cent.
3. Proportion outside tolerance ± 0.0025:
   $$U = \frac{2.0025 - 2.000}{0.001} = 2.5$$

   Proportion outside upper limit = 0.00621 or 0.6 per cent.
   Therefore proportion outside both limits is 1.2 per cent.

14
Estimation and confidence intervals

14.1 Introduction

The process of estimating properties of a population from the evidence furnished in a random sample is known as 'statistical inference'.

There are two steps in the process:

1. Data in a random sample is processed to produce the best single estimate of the required population characteristic. This number is called the 'point estimate'.
2. The point estimate will vary from sample to sample, and will rarely be an exact estimate of the population characteristic. To make the estimate useful, we must be able to assign some limits to the likely deviation of the estimate from the population characteristic. These limits form a 'confidence interval' around the estimate.

For example, we may wish to estimate the mean age of all people in Britain, and a random sample provides a point estimate of 36.4 years; while this is the best estimate we can get from the sample, how close is it to the population mean? Is it accurate to within 6 months, 2 years or what? Within what range would we expect the true value to lie?

Some notations used in statistical inference:

1. Subscript 'p' denotes a population characteristic, e.g., $\bar{x}_p$, σ_p^2.
2. Subscript 's' denotes a sample characteristic, e.g., $\bar{x}_s$, σ_s^2.
3. Point estimates of population characteristics are denoted by a circumflex:
    (a) $\hat{x}_p$ for the point estimate of the population mean;
    (b) $\hat{\sigma}_p^2$ for the point estimate of the population variance.

14.2 Point estimate of the population mean

The best estimate of the population mean is the sample mean. Thus $\bar{x}_s$ is the

INDUSTRIAL ENGINEERING

best point estimate of the population mean $\bar{x}_p$:

$$\hat{x}_p = \bar{x}_s$$

Although individual sample means differ from $\bar{x}_p$ the mean of all the sample means is the population mean.

14.3 Point estimates of population parameters

1. The sample variance σ_s^2 is *not* the best point estimate of the population variance.

   Note: The sample variance is $\sigma_s^2 = \dfrac{\sum(x_c - \bar{x}_s)^2}{n}$ where n is the sample size.

2. The reason for this is as follows: Since $\bar{x}_s$ is used as an estimate of $\bar{x}_p$, the calculation $\sum(x_c - \bar{x}_s)^2$ is always a minimum, and therefore always below $\sum(x_c - \bar{x}_p)^2$.

   Therefore there is a systematic downward bias when σ_s^2 is used to estimate σ_p^2.

   This can be overcome by using a correction factor $n/n-1$ which is demonstrated below.

3. Consider the following population: 4, 5, 6.

   A sample size of 2 is taken and the variance for each sample calculated (Table 14.1).

Table 14.1 Table for calculation of sample variance using correction factor $\dfrac{n}{n-1}$

x_1	x_2	x_s	$\sigma_s^2 = \dfrac{\sum(x_c - \bar{x}_s)^2}{2}$	$\sigma_s^2 \dfrac{n}{n-1}$
4	4	4	0	0
4	5	4.5	0.25	0.5
4	6	5	1	2
5	4	4.5	0.25	0.5
5	5	5	0	0
5	6	5.5	0.25	0.5
6	4	5	1	2
6	5	5.5	0.25	0.5
6	6	6	0	0
$\sum$			3	6

Note: $\bar{x} = 5; \sigma_s = 2/3$;

Mean of $\sigma_s^2 = \dfrac{\sum \sigma_s^2}{9} = \dfrac{3}{9} = \dfrac{1}{3} < \sigma_p^2$.

Mean of $\sigma_s^2 \dfrac{n}{n-1} = \dfrac{\sum \sigma_s^2 \cdot \dfrac{n}{n-1}}{9} = \dfrac{6}{9} = \dfrac{2}{3} = \sigma_p^2$.

Thus we adjust σ_s^2 by $n/n-1$ to remove the downward bias.

The point estimate of the population variance is

$$\hat{\sigma}_p^2 = \sigma_s^2 \dfrac{n}{n-1}.$$

Note 1:
$$\sigma_s^2 \dfrac{n}{n-1} = \dfrac{\sum (x_c - \bar{x}_s)^2}{n} \cdot \dfrac{n}{n-1},$$

$$\therefore \hat{\sigma}_p^2 = \dfrac{\sum (x_c - \bar{x}_s)^2}{n-1}.$$

Note 2: The larger n gets the closer to σ_s^2 becomes σ_p^2.

14.4 Confidence intervals

A confidence interval involves two things:

1. The setting of limits within which the population characteristics can lie.
2. The evaluation of the probability that the population characteristics lie within the set limits; the probability is called a 'confidence level'.

If we set limits, i.e., make the statement that an estimate is within a given distance of the true value, the statement is either true or false; we don't know which. However, if many samples are drawn and the statement is true for 95 per cent of them, then, while we still don't know *with certainty*, the *probability* of being right is high.

14.5 Confidence interval for a mean

This involves two things:

1. Estimating the population mean from the sample mean.
2. Assigning confidence limits. In order to do this we need to know what proportion of the sample means come within a given distance of the population mean, $\bar{x}_p$, i.e., we need to know the *distribution of the sample means*.

From sampling theory we have established the following:

INDUSTRIAL ENGINEERING

1. If n, the sample size, is large (30 or more) then the distribution of sample means closely approximates the *normal distribution* (central limit theorem).
2. The mean of this distribution is the population mean, $\bar{x}_p$.
3. The variance of this distribution, $\sigma_{\bar{x}}^2$, is proportional to the population variance, σ_p^2, and inversely proportional to the sample size, n:

$$\sigma_{\bar{x}}^2 = \frac{\sigma_p^2}{n}$$

4. The proportion of sample whose means lie within any specified distance of the population mean can be found by measuring *the area under the normal curve*.

14.6 An example

A random sample of 442 workers showed a mean working week of 46 hours, and a standard deviation of 7 hours. Estimate the 95 per cent and 99 per cent confidence intervals for the mean working week of all UK workers.

SOLUTION

Step 1 From the sample calculate

$$\bar{x}_s, \sigma_s^2.$$

Step 2 Make point estimates

$$\bar{x}_p, \sigma_s^2$$

Point estimate of

$$\bar{x}_p = \bar{x}_s$$

Point estimate of

$$\sigma_p^2, \hat{\sigma}_p^2 = \sigma_s^2 \cdot \frac{n}{n-1}$$

$$\therefore \hat{\sigma}_p^2 = \frac{7^2 \cdot 442}{441}$$

$$= \frac{49 \times 442}{441}$$

Step 3 Calculate the standard deviation of the distribution of the sample mean

$$\sigma_{\bar{x}} = \sqrt{\frac{\sigma_p^2}{n}}$$

ESTIMATION AND CONFIDENCE INTERVALS

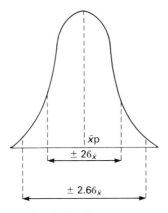

Figure 14.1 Confidence intervals calculated

Since we don't know σ_p^2, we use its *point estimate* $\hat{\sigma}_p^2$.

$$\therefore \sigma_{\bar{x}} = \sqrt{\frac{\hat{\sigma}_p^2}{n}}$$

$$= \sqrt{\frac{49 \cdot 442}{441} \cdot \frac{1}{442}}$$

$$= \sqrt{\frac{49}{441}} = \frac{7}{21}$$

$$\sigma_{\bar{x}} = 1/3 \, h$$

Step 4 From the properties of the normal distribution we can calculate the confidence intervals (Fig. 14.1).

The 95 per cent confidence intervals are:

$$\bar{x}_s \pm 2\sigma_{\bar{x}} = 46 \pm 2 \cdot \tfrac{1}{3}$$

$$= 46 \pm \tfrac{2}{3}$$

$$\therefore \text{Range} = 45\tfrac{1}{3} - 46\tfrac{2}{3} \text{ hours}$$

The 99 per cent confidence intervals are:

$$\bar{x}_s \pm 2.6\sigma_{\bar{x}}$$

$$= 46 \pm 2.6 \times \tfrac{1}{3}$$

$$= 46 \pm 0.87$$

Range = 45.13 to 46.87 hours

INDUSTRIAL ENGINEERING

SOME IMPORTANT NOTES ON CONFIDENCE INTERVALS

1. The general method for estimating the confidence interval for the population mean is given by the following formulae:

$$\bar{x}_p = \bar{x}_s \pm 2\sigma_{\bar{x}} \tag{1}$$

$$\bar{x}_p = \bar{x}_s \pm \frac{2\sigma_p}{\sqrt{n}} \tag{2}$$

2. The larger the sample size n the smaller the interval at a given confidence level, i.e. the more accurate the estimate of $\bar{x}_p$.

14.7 Another example

Frequently, we want to know *what size of sample must be drawn* to get within a specified distance at a given confidence level, e.g., we want to know the mean working of UK workers to within half an hour either way, at 95 per cent confidence level.

Step 1 From *given level of accuracy*, calculate the standard deviation of the distribution of sample means using the following relationship:

$$\bar{x}_p = \bar{x}_s \pm 2\sigma_{\bar{x}}$$

At 95 per cent confidence level:

$$\bar{x}_p = \bar{x}_s \pm 2\sigma_{\bar{x}}$$

But

$$\bar{x}_p - \bar{x}_s = \pm \tfrac{1}{2} \text{ hour}$$

$$\therefore 2\sigma_{\bar{x}} = \frac{1}{12} \quad \text{or} \quad \sigma_{\bar{x}} = \tfrac{1}{4} \text{ hour}$$

Step 2 Calculate the value of the sample size required n, which yields the accepted standard error calculated in step 1 using the relationship below:

$$\sigma_{\bar{x}}^2 = \frac{\sigma_p^2}{n}$$

$$\frac{1}{16} = \frac{\sigma_p^2}{n}$$

$$n = 16\sigma_p^2$$

Step 3 Since σ_p^2 is not known, estimate its value by conducting a pilot study. For purposes of this example let us use the results of the example, given in section 14.6 as the results of our pilot study, i.e.,

$$n = 442 \text{ workers} \left.\begin{array}{l} \\ \\ \bar{x}_s = 46 \text{ hours per week} \\ \\ \sigma_s = 7 \text{ hours} \end{array}\right\} \begin{array}{l} \hat{\sigma}_p^2 = \sigma_s^2 \cdot \dfrac{n}{n-1} \\ \\ \phantom{\hat{\sigma}_p^2} = 49 \cdot \dfrac{442}{441} \\ \\ \hat{\sigma}_p^2 = 49 \text{ (approx.)} \end{array}$$

Step 4 Calculate the sample size required using the point estimate for the population variance, calculated above.

$$n = 16\sigma_p^2 \quad \text{or} \quad (16\hat{\sigma}_p^2)$$
$$= 16 \times 49 = 784$$
$$= \underline{\text{say } 800}$$

A sample size of approximately 800 workers is required for a 95 per cent confidence interval of $\pm\frac{1}{2}$ hour.

15

Operational research

15.1 Introduction

Until the second half of the nineteenth century, most enterprises were owned and managed by one individual. He was able, by virtue of the limited size of the enterprise, to control directly all the activities within it, from buying material through processing to selling the product. However, with mechanization and the resulting division of manual labour came the division of managerial labour. This division of management was by function—financial, production, marketing, personnel, etc. A little later there occurred further segmentations when the growing enterprise set up factories in different locations.

Where, before, technology had been mainly responsible for advances in industry, early in the twentieth century there was increasing concentration on improving the effectiveness of labour. Psychology was applied to motivate the operators and work study and industrial engineering were used to design methods and procedures. In the marketing field, too, statistics, economics, sociology and psychology were all brought to bear more and more.

During this development of management techniques there was one conspicuous gap: science did not aid the *executive* function created by the division of management. The executive function may be defined as follows:

> To integrate the policies and operations of all the diverse departments reporting to the executive in order to obtain the overall operation that comes as close as possible to realizing the organization's overall objectives.

The task of the executive is complex because of the conflict of interests. Each of the management functions has its own criteria for success. The objectives of some of the major functions may be summarized as follows:

- *Production*   To maximize output and minimize unit costs.
- *Marketing*   To maximize sales volume and minimize unit cost of sales.
- *Finance*   To minimize capital required to operate the business.
- *Personnel*   To maximize employee morale and minimize labour turnover.

These separate objectives are sometimes inconsistent. Consider the problem of inventory policy.

Production likes long continuous runs to minimize setup costs and to gain efficiency by long practice. This requires large inventory and few products. Marketing likes short delivery and plenty of product choice. This requires large inventory and *many* products. Finance likes small inventories during slack periods—it may allow inventory to grow somewhat during busy periods. Personnel likes to maintain production when sales drop to retain skilled labour and to maintain morale—i.e., in conflict with financial objectives, i.e., large inventory of personnel.

The executive has to optimize the overall effect. The serious work on this type of problem had to wait until the end of the Second World War. The reasons were as follows:

1. Before the Second World War, those scientific brains which were employed in industry were concentrated on technological research in such a way that little contact if any was made with the problems of operation.
2. The problems of operation were made urgent by nationalization immediately after the war.
3. The problems of strategy during the war were exactly analogous to those encountered in industry and the best scientific brains were brought to bear on these problems.

The field of activity developed to deal with these problems of strategy and optimization is called operational research—sometimes operations research or operational analysis.

15.2 Definition and activities

Operational research (OR) is a scientific method for providing executive departments with a quantitative basis for decisions regarding the operations under their control. OR is an applied science using all known scientific techniques as tools in solving a specific problem. It uses mathematics but is not a branch of mathematics. It uses the results of method study but is not efficiency engineering. It is not a branch of engineering—the engineer is a consultant to the builder of equipment while the OR worker is a consultant to the *user* of the equipment.

Operational research is quantitative. Certain aspects of practically every operation can be measured and compared with similar aspects of other operations. It is these aspects which can be studied scientifically. Many problems apparently very different can be shown to be different only in content. Their form may be similar in many respects. To recognize and isolate these similarities is a major part of OR. The quantitative aspects of the problems are not the whole story in most executive decisions. Many others enter—politics, morale, industrial relations, personnel relations. These must be added to the quantitative basis provided by OR to reach the final decision.

The presentation of results in a precise, clear form understandable by the executive is most important.

Operational research is a staff function which should be in direct contact with the executive. It should *not* report to the research department.

15.3 Outline of problem forms

The basic features of the OR approach is to set up a mathematical model and to identify the form of the problem. Sometimes the solution will be obtained by accounting for more than one form within the same problem.

There are a number of variations on each form, and indeed there is no generally accepted unique classification of problems. However, eight forms, singly or in combination, account for most executive problems. They are as follows:

- Inventory
- Allocation
- Queuing
- Sequencing
- Routing
- Replacement
- Competition
- Search

INVENTORY PROBLEMS

Inventory is any idle resource. The inventory problem is one of optimization of overall cost in the face of two types of cost:

1. Costs that increase as inventory increases.
2. Costs that decrease as inventory increases.

The problem is to select the quantity or frequency of acquisition so that the overall cost is minimized.

Inventory may include such resources as plant, capacity in terms of plant equipment, labour, raw material, finished product, semi-finished product, floor space, capital, etc.

Mathematical techniques for attacking inventory problems are highly developed. They involve use of calculus and probability theory, also matrix alegbra and calculus of variations.

ALLOCATION PROBLEMS

The simplest category of allocation problem is defined by the following conditions:

1. There is a set of jobs of any type to be done.
2. Enough resources are available for doing all of them.
3. At least some of the jobs can be done in different ways—by using different amounts and combinations of resources.
4. Some of the ways of doing these jobs are better than others, e.g., are less costly.
5. There are not enough resources available, however, to do each job in the best way.

The problem is to allocate jobs to resources in such a way that the overall cost is a minimum or overall profit a maximum.

The second category of allocation problem arises when there are more jobs to be done than available resources permit. The question of selection of jobs to give maximum returns is added to the question of allocation. The familiar 'product-mix problem' is in this category.

The third type of allocation problem is that of determining what resources should be added or subtracted, i.e., where there is control over resources.

Most of the techniques used to solve allocation problems are under the heading 'mathematical programming'.

QUEUING PROBLEMS

These arise where units arrive at some facility which serves and eventually releases them. Schematically the system may be presented as shown in Fig. 15.1.

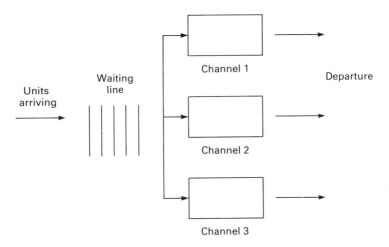

Figure 15.1   Queuing system

INDUSTRIAL ENGINEERING

There are three basic elements in the system: (1) the time distribution of arrivals, (2) the service mechanism (number of channels and distribution of service time), (3) the queue discipline (order in which units are served). Units can mean customers or jobs.

It is apparent that some inventory problems can also be treated as queuing problems. Items in stock are service facilities waiting for customers. When the stock is exhausted a queue of customers is the result. The mathematical theory is highly developed and simulation, including Monte Carlo methods, is used where the mathematics is too complex.

SEQUENCING PROBLEMS

This type of problem is concerned with selecting a queue discipline so that some appropriate measure of performance is maximized—minimum time or cost, etc. Where there are only a few operations (or customers) the problem is easily solved by inspection. Where there are many jobs and channels we have to experiment using a computer and even here only approximate solutions can be obtained.

A widely used technique to handle the sequencing problem in planning a large complex project is that of network analysis (or critical path analysis).

ROUTING PROBLEMS

The best known of these is the 'travelling salesman' problem. The salesman has a number of locations he has to visit, starting and finishing at base. The problem is to select a route going through each location only once, in the shortest possible distance (or time or cost). There are two types of problem, as follows:

1. The symmetrical, where the cost, distance, time, etc., between any two locations A and B is the same whichever the direction of travel.
2. The asymmetrical, where the cost etc. depends on the direction; for example, there may be hills involved in the route.

A general solution by mathematical analysis has not yet been worked out; however, numerical methods have been found.

The problem of scheduling production runs on different products where setup (and takedown) costs are incurred when changing over from one to another is in fact a routing problem of the asymmetrical kind.

REPLACEMENT PROBLEMS

These are of the form, 'What is the optimum point in time to replace resources of limited life?' Two types are considered: those involving items that degenerate with use and those which die or fail utterly after a time.

There are two costs to be reconciled in general. The cost of maintenance (and production) increases the longer a deteriorating piece of equipment is retained, while the cost of frequent replacement raises investment costs.

Dynamic programming is the technique most generally applicable to the first type of problem. Analytical simulative techniques are used for the second type; in particular, knowledge of the distribution times to failure is essential to the solution.

Scrap allowance problems come under the heading of replacement problems.

COMPETITIVE PROBLEMS

Here the executive decision is affected by decisions made by one or more other factions in competition. Three classes of situation may be described, namely:

1. The competitors' actions are known in advance without error.
2. Competitors' reactions can be predicted but the predictions are subject to error.
3. Nothing is known about what competitors will or are likely to do.

Theory of games has developed general ideas about competitive situations but can only provide solutions for simple problems. Simulation is valuable by setting up the context of the competition and allowing real decision-makers to operate with it. This technique is called 'gaming'.

SEARCH PROBLEMS

All problems of estimation and forecasting are search problems. The problem may be to design a method of looking most effectively for something specified in a continuum (of time or of space). Costs have to be reconciled, i.e., those involved in the search reconciled with the cost of missing what is looked for coupled with the risk of missing it.

Two kinds of error can be made: (1) failure to detect what one is looking for because of inadequate coverage ('sampling' errors), and (2) failure to detect what one is looking for even though one has looked in the right place, or erroneous 'detection' of the thing which is not there ('observational' errors).

The larger the sample the less likely is sampling error, but the less time spent per observation the more likely is observational error to occur.

The following are all search problems: auditing; mineral exploration (prospecting); design of quality control techniques. The best location of goods

INDUSTRIAL ENGINEERING

in a supermarket or department store is a search problem in reverse. The condensation, storage and retrieval of information, particularly scientific, is a search problem.

SUMMARY

Many problems are found to be similar in their form, although they are at first apparently unrelated. When this is so, some general method of treatment is usually appropriate. Even if mathematical analysis is not possible either numerical or simulation methods can often yield solutions.

In general, a problem will contain a mixture of forms.

15.4 Essential characteristics of OR

SYSTEMS ORIENTATION

The activity of any part of an organization has some effect on the activity of every other part. To evaluate any decision, therefore, it is necessary to identify all the significant effects and to evaluate their combined effect on the performance of the organization as a whole.

The approach is not to simplify the problem by ignoring the factors that made it difficult to solve: rather, the problem is expanded until all the significantly interacting components are contained within it. In many instances, this approach provides the manager with a basis for initiating inquiries at a higher level into policies which seem to be affecting his own performance adversely.

INTERDISCIPLINARY TEAMS

It is recognized more and more that it is necessary to look at a problem from as many angles as possible. In the past, disciplines and studies have been classified and compartmented, not least scientific studies. This has been accepted for so long that we have been in danger of believing that problems are similarly compartmented by nature, e.g., 'a chemical problem', 'a biological problem', etc. This clearly is nonsense. It follows that the best solution (there are usually many alternative solutions to practical problems) may be obtained by using not one discipline, but by using many, or more than one.

OR METHOD

OR method is that of scientific research. This is essentially experimental. Opportunities for experimentation in the natural environment are few owing to the characteristics of the system under study. Recourse must therefore be had to what is called a 'mathematical model'. Experiments are made on this in abstract terms by manipulating the variables.

The basic form of all OR models is where the measure of the system's overall performance is a function of both controlled and uncontrolled aspects of the system. This may be expressed symbolically as follows:

$$P = f(C_i, U_j)$$

where P is a measure of the system's overall performance
C_i is a set of controlled aspects of the system
U_j is a set of uncontrolled aspects of the system

The solution yielded consists of one equation for each controllable variable of the form:

$$C_i = f_i(U_j)$$
$$C_2 = f_2(U_j)$$
$$C_3 = f_3(U_j) \text{ etc.}$$

These equations are called 'decision rules'.

Part 5

Industrial engineering on the shop floor

16

Shop floor data collection

16.1 Introduction

It is becoming increasingly clear that what happens on the shop floor has a significant effect on the performance of a manufacturing company. Quite often, elaborate plans and preparations are made, only for it to be found that at the place they really count, on the shop floor, they really go awry. Plans are very often made by the professional production controller, often aided by a computer, but tend to be implemented at the level of foremen or supervisors who coordinate and control between themselves what really happens. Frequently the two levels do not meet, i.e., the plans and results are very different.

THE SHOP FLOOR'S IMPORTANCE

The importance of shop floor performance, particularly in the small-lot manufacturing company—which is the prevalent type of factory in the United Kingdom—is increasingly becoming apparent. The shop floor is a dynamic area in which the workforce, machines and equipment, materials and tools are coordinated by dispatching, routing and expediting of tasks, by reacting to schedules, priorities and set-ups, and reporting order/job and resource status.

Changes are frequently required of the operator owing to any of the following:

1. Method changes due to design modifications and sometimes to poor design.
2. Production plans being altered, sometimes because 'someone special' needs his order before everyone else!
3. Quality standards which have to be tightened or temporarily overcome.
4. Rework due to own or others' fault.
5. Routing alterations.
6. Rescheduling.
7. Equipment failure.

INDUSTRIAL ENGINEERING

8. Material delays and shortages.
9. Priority changes.
10. Tool breakage.

It is on the shop floor that our plans and preparations become practical realities. The need for order and control is vital. Shop performance needs to be understood if we are to succeed. Measurements, information, improvement and a good knowledge of people become necessary.

THE KEYS TO CONTROL

What are the key ingredients for shop floor control—control in the sense of achieving objectives rather than making authoritarian demands of people? They are:

1. Good layout of facilities, with proper, work-designed methods of operating.
2. Well-designed physical flow of materials to and from the workplace. Untidiness is a symptom of laxness.
3. People selected and trained for their tasks and who take part in teamworking exercises to inspire themselves, e.g., quality circles.
4. Superiors who also are selected and trained for their tasks.
5. Good production planning and control, systems and people.
6. Accurate systems for the collection and transmission of data.
7. Good equipment and work aids.

16.2 Data collection

Shop floor control, whether it is by all or by any of the ingredients listed above, depends upon accurate and timely feedback information being available. In the past, this data was collected manually. The vast majority of companies still rely on this particular style of information flow. The larger companies, and increasingly the medium-sized ones, are now in many instances using microprocessor-based terminals to capture this same information.

MANUAL SYSTEMS

These comprise any of the following:

– Work or job tickets
– Clock cards
– Route cards
– Transit and transport cards
– Operator timesheets

SHOP FLOOR DATA COLLECTION

INFORMATION REPORTS

Information coming from the shop floor is needed for a variety of different tasks and functions. Some examples are:

1. Attendance time: Wages office, personnel, costing.
2. Output records: Production control, supervisors and managers. Costing, wages office (incentives).
3. Lost time, waiting time: Industrial engineering, supervisors, managers.
4. Transit of work: Production control, supervisors and managers.
5. Breakdowns: Maintenance, production control, supervisors and managers.

Factory data therefore needs to be captured, stored and processed for manufacturing systems that serve a variety of needs. Modern systems require immediacy and data from the source of activity. Accuracy, punctuality and completeness are important considerations in recording shop floor status.

16.3 Data capture by computers

There are generally three main forms of data capture by computers:

1. Operator-controlled terminals.
2. Equipment-monitoring devices.
3. Time and attendance terminals.

OPERATOR-CONTROLLED TERMINALS

These are defined as microprocessor-controlled devices distributed on the shop floor with transmission links to a computer or controller, which are input by means of a simplified keyboard or preassigned data recorded on a badge, card or document, or automatic signal pickup, such as weight, quantity time, or bar code.

Forms of data gathered

The typical forms of data gathered are:

– Time and attendance (integrated into the data collection terminal).
– Job status.
– Stores recording.
– Material location.

- Tool location.
- Inspection reporting.
- Equipment status.
- Operator status.

The terminals are usually robust to cope with shop floor conditions such as temperature, dirt, oil, fumes, vibration, noise, humidity, liquids and some kinds of electromagnetic interference. Stationery terminals are arranged on the shop floor so that perhaps 20 to 30 operators can record their transactions near their place of work. Time and attendance terminals are usually placed near entrances/exits and can cater for much greater numbers of people.

Forms of input

Typical of input are the following forms:

Numeric keyboards with other function keys

These comprise the numbers 1 to 9, zero, decimal, plus or minus, and other special keys, such as enter, send, cancel, clear line, hold; message types can be coded as with the following:

1. Attendance.
2. Work in progress.
3. Waiting time.
4. Goods in.
5. Inspection.
6. Rework.
7. Stores issue.
8. Team makeup.
9. Overtime authorization.
10. Exception.

Sub-codes can also be elaborated, e.g., types of waiting time.

Prepunched cards

These are more prevalent with the older types of system, 80-column punched cards having almost disappeared from the data processing environment.

Identity (ID) cards

These are normally heat-sealed in a plastic cover and can contain photograph, name and bar code for employee number. The bar code can also be masked to prevent illegal copying.

SHOP FLOOR DATA COLLECTION

Alphanumeric keyboards

Similar to numeric keyboards but with the range of alphabet characters included. These types can be portable, being connected via communication link to the computer.

Card-encoded data

The data would normally be encoded by magnetic strip and the cards would be inserted into a slot on the terminal to record, e.g., job number.

Bar-coded data

This type of input is becoming increasingly popular. The goods or items each have a bar-coded label attached which is read by light pens, hand-held guns or wipe-through reading devices. For instance, goods entering or leaving a warehouse can be scanned and used to update stock records.

Analogue or digital input

Analogue input could for instance be temperature, pressure and humidity read directly from instruments or sensors. Digital input is where data is digital and is then sent to the computer. A typical use of such information is with automatic weighing equipment employing electronic counting. The count can be calculated by the computer and then used to update stock records or production control systems.

Other types

These would include optical character recognition (OCR), magnetic character recognition (MCR), inductive, reflective tape, or infra-red. Voice recognition systems are not yet commercially available, but microprocessor chips are available for pre-registered voice patterns.

Forms of output

Types of output provided from the terminals can be:

Alphanumeric displays

These can be provided by a VDU screen, but more often consist of an LED display on a flat screen.

INDUSTRIAL ENGINEERING

Sound signals

These are often given to let the operator know that a signal has been transmitted, for instance on pressing the 'send' button.

Controls for other equipment

The output can be used to control other equipment, through relaying it by means of digital or analogue signals. In this case the terminals would be operating in a manner similar to that of PLCs (programmable logic controllers), though they are not principally for this type of process control operation. Energy management could fall within this category.

Voice response

Microprocessor chips for voice response are now commercially available, and are much easier to install than voice recognition systems. There are various ways in which sounds can be simulated by electronic signals, called 'voice synthesisis'.

Terminals are often programmable through simple defined statements, so that variations in their operations are possible. For instance, one particular type has 8K bytes of EPROM and 8K bytes of CMOS RAM using digital Tll, 16-bit microprocessors. These can also be programmed downline from a host computer. Internal batteries can also provide emergency power supplies when needed.

Portable data terminals are available with solid-state or cassette memory to store recorded transactions. These can be used continuously for up to 10 hours with an internal rechargeable battery. Entry is normally by simple keyboard or optical wand, with transmission by inserting a plug into a socket, or through an acoustic coupler on a telephone line.

EQUIPMENT MONITORING DEVICES

These are microprocessor-based terminals which directly monitor the state of a process and communicate this through transmission-links to a host computer or controller. A device can be completely automatic or can be partly automatic and partly operated by an operator.

Such devices are used primarily to measure the utilization or running time of machines and equipment, although they can be used to monitor other physical attributes, where it is more normal to use PLCs or process control equipment than it is to use equipment monitoring devices.

The device would normally monitor activity or equipment status using sensors. These can be lasers, photoelectric cells, microswitches or any other measuring devices or instruments from which a digitized signal can be encoded. They are used to record each rotation of a shift in a machine, each stroke of a reciprocating motion or when a device is idle or broken down. They are not so much condition-monitoring devices, e.g., relaying through sensors why a machine is broken down or circuit analysers for electronic equipment. The status of the equipment is normally transmitted to a local computer and indicator lights, printer logs or screen digital displays provide details of current status or perhaps a daily log.

Linked to the automatic monitoring in many systems is a manually operated keyboard for providing status information, typical examples being as follows:

Status Reason
- 0   Setting up
- 1   Lack of orders
- 2   Lack of materials
- 3   Raw material difficulties
- 4   Tool change
- 5   Tool repair
- 6   Breakdown of equipment
- 7   Repairs
- 8   Maintenance (planned)
- 9   Break

In one such device, the Kienzle Machine Terminal 2450, a card records the time spent producing and time for each of those status conditions with a printout on the card of quantity produced, times, and utilization percentage automatically calculated at the end of each day or shift.

Other machine interface terminals can be used on the following:

1. NC to CNC upgrade.
2. Two-way part programming transfer to machine.
3. Part programming and editing facilities.
4. General-purpose interface (as PLC).

Many types of industries have used such devices, e.g.,

- Engineering
- Electronics
- Food
- Cosmetics
- Plastics
- Aerospace
- Materials handling
- Paper
- Machine tools

INDUSTRIAL ENGINEERING

TIME AND ATTENDANCE TERMINALS

These devices are principally used to record employees' time and attendance. Some systems will collect clocking data, record it, process it, and produce reports and will update other computer systems, e.g., payroll and costing. They can cope with various working patterns, e.g., fixed hours, flexible working hours, fixed or rotary shift patterns and core time, and with a simple entry record holidays, sickness and other absences. A system will also distinguish between time for normal hours and time for various overtime rates of pay. Employees are given a plastic card—like an ordinary credit card—which is inserted into the terminal, which they keep on their person. Clock cards and card racks are therefore eliminated.

Outputs can vary, but can include:

- Employee name, number and attendance details
- Unapproved overtime list
- Absentee list
- Departmental list
- Holiday and sickness list

SUMMARY

With modern MRPII closed-loop manufacturing systems there will be an increasing need to have accurate and quick shop floor control information. This is one of the keys to higher productivity and quality performance for manufacturing in a world market.

Factory authorization of equipment must be expanded to include control and decision-making functions and shop floor terminals are going to provide the basic data for this type of control in the future.

Most of the perhaps larger and more organized companies have already used the equipment described, but there is evidence that medium-sized companies can now embark on the feasibility for such an exercise.

Shop floor control, though perhaps unglamorous, is the key to productivity and quality performance.

16.4 Suppliers of equipment

OPERATOR-CONTROLLED TERMINALS

Bedat ATS UK Limited, 80 City Road, London EC1Y 2AR. Tel. 01-253 8102.

FIS Kins Applied Technology Ltd, Kins House, 141 Garth Road, Morden, Surrey EM4 4LF. Tel. 01-330 6111.

SHOP FLOOR DATA COLLECTION

8037 Triad Computing Systems Ltd, 42 Kingsway, London WC2B 6EX. Tel. 01-831 7211.

Shop floor Kewill Systems plc, Ashley House, 20–32 Church Street, Walton-on-Thames, Surrey KT12 2QT. Tel. (0932) 248328.

480 Feedback Data Ltd, Bell Lane, Uckfield, East Sussex TN22 1PT. Tel. (0825) 4222.

Mandate Istel Ltd, PO Box 5, Grosvenor House, Redditch, Worcestershire B97 4DQ. Tel. 0527 64274.

MACHINE MONITORING

Datascan SPL Industrial Division Battersea Road, Healton Mersey Stockport SK4 3EA Tel. 051-442 9552

Rhombus Rhombus System Ltd, 24 Downham's Lane, Milton Road, Cambridge CB43 1XT. Tel. 0223 356986.

MIT Tangram, Computer Aided Engineering Ltd, 5 Sidderlay Way, Royal Oak Industrial Estate, Daventry, Northants NN11 5PA. Tel. 0327-705026.

2450 Kienzle Data Systems Ltd, 224 Bath Road, Slough SL1 4DS. Tel. 0753-33355.

A large number of electronics/process-control computer companies will provide PLCs linked to other computers.

TIME AND ATTENDANCE

Tempus ATS UK Ltd, 80 City Road, London EC1Y 2AR. Tel. 01-253 8102.

8039, 495, 496/2 Triad Computing Systems Ltd, 42 Kingsway, London WC2B 6EX. Tel.

Mandate Istel Ltd, PO Box 5, Grosvenor House, Redditch, Worcestershire B97 4DQ. Tel. 0527 64274.

OTHER MAJOR SUPPLIERS OF ALL SYSTEMS

IBM; NCR; Honeywell; ICL; Hewlett Packard.

17

Production scheduling of multistage and jobbing production units

17.1 Definitions

A 'job' is defined as a complete order needing time on several of the machines (or work centres) in the workshop.

A 'task' is defined as an individual operation on one of the machines, i.e., a job is a set of tasks.

'Scheduling' is the process of assigning starting and finishing times on each job on each machine. In basic terms, this can be performed manually using a Gantt chart (Fig. 17.1), where the sequence of operations for each order is loaded against each facility on a chart with a horizontal time-scale, usually in hours or days. It is simply a visual aid for the manager, who can quickly see the interaction of jobs and facilities against time. There are many adaptations to the Gantt chart, such as loading boards, magnetic boards, marker boards and so on. Some of these are proprietary boards offered for sale by companies for scheduling purposes.

The term 'sequence' often refers to the situation where a queue of known jobs are arranged in order and planned through a series of discrete facilities (one after the other in a given order, e.g., in food processing and canning).

'Dispatching' usually relates to a situation where jobs arrive continuously in a queue for processing on only one machine. It generally follows some form of aggregate planning, then capacity planning for each facility, then the delegation of sequencing to machines or people or combinations, using dispatching rules.

A 'dispatching rule' or 'priority' chooses what task to perform next from the set of tasks waiting at a given machine.

The prediction of finishing times for jobs and the accuracy with which these predictions can be met has long been a very difficult problem for management to resolve. Scheduling systems have been used to help in this area.

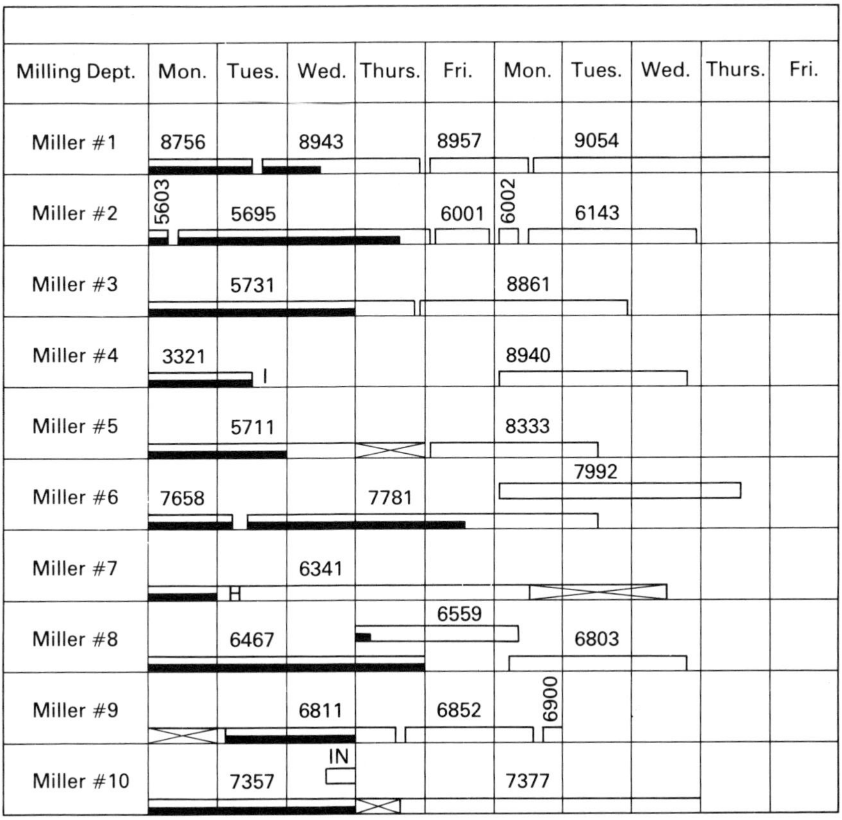

PRODUCTION SCHEDULING

Milling Dept.	Mon.	Tues.	Wed.	Thurs.	Fri.	Mon.	Tues.	Wed.	Thurs.	Fri.
Miller #1	8756		8943		8957		9054			
Miller #2	5603	5695			6001	6002	6143			
Miller #3		5731				8861				
Miller #4	3321					8940				
Miller #5		5711				8333				
Miller #6	7658			7781			7992			
Miller #7			6341							
Miller #8		6467			6559		6803			
Miller #9			6811		6852	6900				
Miller #10		7357	IN			7377				

Figure 17.1 Gantt chart

A job shop can be envisaged as a set of queues of jobs waiting to be processed. Once finished, a job joins another queue at the next process, until the tasks are completed.

The sequence and flow of one order may be quite different from those of another and so process flows (how machines are arranged on the shop floor) and material flows are likely to be at cross purposes, unless production flow analysis and relayout have been accomplished, in which case the majority of the process and material flows may be the same, but not all.

17.2 Dispatching rules

Shop scheduling using dispatching rules has no easily applied solution. It can often be thought of in terms of a complex of differing needs and requirements which have to be balanced, planned and executed by human ingenuity, sometimes with the aid of a computer.

Instances of some factors which have to be considered are as follows:

1. Production will often be made up of differing batches. The economic batch quantity may not be the same as the movement batch quantity. Large batches of work may be mixed with small ones.
2. There may be a wide range of parts, components and materials to be considered and the flow of these will be from different sources, e.g., steel bar from the stores, castings from the foundry, parts from the previous operation.
3. Orders could be mixed, some being made to order with a promise date to the customer, others made to stock, perhaps without a promise date but with a requirement to be made before stocks are used up.
4. The planned sequence operations are likely to be different for each part, and some of these are likely to be 'outside' processes, e.g., heat treatment or finishing operations which are offloaded to another plant. There are therefore unique routes through facilities.
5. There will be a variety of due dates for different jobs.
6. Material availability and tool availability have to be linked to process facility and operator manning patterns.

With dispatching there will need to be a balance between conflicting objectives:

1. Producing orders on time, or with minimum lateness, in order to give the customer a reliable delivery service and a reasonable lead time for his order.
2. Minimizing the average throughput time of jobs, thus reducing the value of work in progress and hence money tied up in inventory. The way in which processes are arranged physically on the shop floor can often satisfy this criterion, e.g., group technology, production cells.
3. Achieving maximum utilization of equipment or people. There is an increasing trend towards more sophisticated and expensive but adaptable machinery, e.g., computer numerical control (CNC) equipment. This needs to be planned much more carefully and thoroughly. Cost per machine may be higher, process time drastically reduced and batch quantities may be smaller as setting time is reduced (automatic tool changing).

A survey of dispatching rules for manufacturing job shop operations was reported by researchers Blackstone, Phillip and Hogg (*International Journal of Production Research*, 1972) and by Conway, Maxwell and Miller, *Theory of Scheduling* (Addison-Wesley, 1967), and by many others. See also chapters in Lockyer, *Production Control in Practice* (Pitman, 1975).

PRODUCTION SCHEDULING

Some possible dispatching rules are the following:

1. *SOT* (*shortest operation time*)   Select from among those waiting the task with the shortest operation time on the machine.
2. *FCFS* (*first come first served*)   Select from among those waiting the first job that arrived at the work centre.
3. *TSOT* (*truncated shortest operation time*)   This is an adaptation of rule 1: select according to rule 1 (SOT) unless a specified, truncation time has elapsed, in which case that job goes to the front of the queue.
4. *DS/RO* (*dynamic slack per remaining operation*)   Calculate the relative priorities of the queued tasks as follows:

$$\frac{\text{Machine available time until due} - \text{total process time remaining}}{\text{number of operations to be completed}}$$

Tasks with the highest results are performed first.

5. *CR* (*critical ratios*)   Calculate the priorities as follows:

$$\frac{\text{Machine available time until due}}{\text{estimated lead time (i.e., processing} + \text{transfer} + \text{waiting)}}$$

Tasks with the lowest results are chosen first. There can be many forms of this rule because lead time will depend upon the shop load in relation to capacity.

6. *COVERT* (*cost over time*)   Calculate the priorities as follows:

$$\frac{\text{Expected cost of tardiness}}{\text{operation time}}$$

The tasks with the highest results are performed first.

These formulas are given in Baker, *Introduction to Sequencing and Scheduling* (Wiley, 1974).

The rule which seems to provide the best result, from among a whole series of simulated research projects, is the SOT rule. Because the shortest jobs are run first, the larger number are completed earlier and the larger jobs are held back the most. Note that the larger jobs are often the more profitable ones in terms of 'contribution to profit' and Pareto analysis, in which case they may be the more important ones!

If work load on the shop is tight (over 80 per cent loaded) then SOT works better. If work load is loose (less than 80 per cent loaded) then DS/RO, CR and COVERT perform well (Blackstone *et al.*).

17.3   Defining service levels

In general, it is usual to find that the work load requirement in a service

INDUSTRIAL ENGINEERING

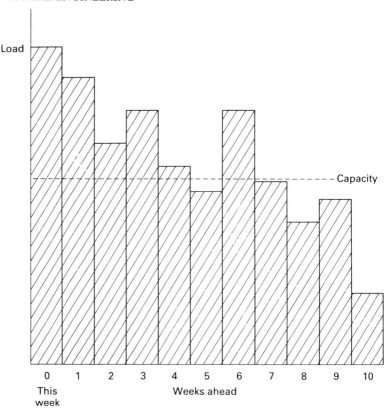

Figure 17.2 Work load diagram

department (e.g., a machine shop, or a fabrication shop serving assembly) has overloads for the first few weeks, tailing down to a smaller load in later weeks; see Fig. 17.2. The 'this week' load would normally include those orders planned for this week and backlog. This is very typical of what is found in industry—most things are needed yesterday!

If the shop is a department providing a service (either to assembly or to the customer) then the relationship between capacity load and service levels will apply; see Fig. 17.3.

'Service level' can mean either of the following:

1. The service level from stocks (make-to-stock situation), e.g., the percentage number of requests from stocks which were met in a given period.
2. The ratio of on-time deliveries to deliveries promised in a given period.

Each company will have to decide its own rules for service level, degree of

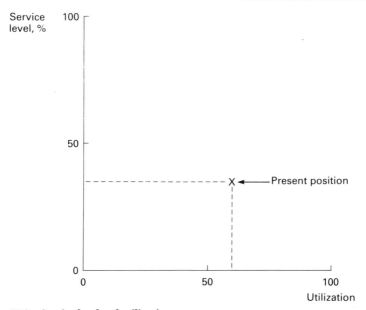

Figure 17.3 Service level and utilization

lateness, etc. Generally, UK manufacturing industry does not provide a very rosy picture regarding delivery performance. See New and Sweeney, 'Delivery performance and throughput efficiency in UK manufacturing industry', *Journal of Physical Distribution and Materials Management*, Vol. 14, No. 7, 1984.

'Utilization' is considered to be the time when the resource is used for performing useful work, divided by the time available.

If in a given company the service level is measured and found to be 35 per cent while the overall utilization of the facilities is 60 per cent then it will appear in Fig. 17.3 above as 'present position' X. To improve service level to a higher percentage, we could bring in more facilities and so reduce utilization. We might for instance obtain 50 per cent service level for 50 per cent overall utilization. We could also increase the utilization of existing resources, say to 80 per cent and subsequently reduce the service level to 20 per cent.

By better planning, scheduling and control, it would be possible to increase both service and utilization, but not to 100 per cent in each case. This could be achieved only with a known, cyclical and balanced flow of work, where all circumstances are known and planned for. Job shops however are characterized by chaos, with jobs running late and large quantities of work in progress, inventory congesting work areas, and schedules that are constantly being rearranged by managers superimposing their priorities or operators and supervisors doing their own thing.

INDUSTRIAL ENGINEERING

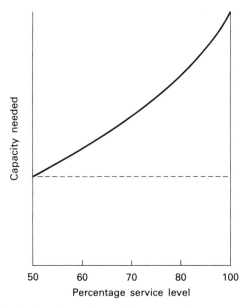

Figure 17.4 Service level and capacity needed

Normal-service-level mathematical criteria can be applied to production capacities. If capacity is balanced to average demand (the mean of a series of loads being related against capacity) we must expect only a 50 per cent service level. If we measure the standard deviation of the loads, then it is possible to relate k (a constant) times the standard deviation to the service level needed. A graph can then be calculated (Fig. 17.4).

It can be seen why service departments do not provide the service level normally expected of them. Industries with jobbing shop production usually overcome the problem of service level by varying the delivery time for work. The larger the order book, the longer it will take for the customer to get his goods.

There can be economic sense in providing more facilities, costing, say, £100 000 per annum if it will produce an additional throughput of, say, £500 000 per annum. An extra facility in an overloaded department can often release very expensive assemblies which are being held up and are of much greater value than the additional cost.

Scheduling can be effective only if there is sufficient capacity.

17.4 Keys to scheduling

To overcome the many practical problems experienced in industry, the following rules are important.

SETTING REALISTIC DELIVERY DATES

Many engineering companies plan one week for each operation to be completed. Studies have shown that rules such as these are not realized in practice, even by the best companies. In one particular study undertaken by PE consultants the best was 1.25 weeks per operation and the worst much greater.

Each company should establish its own critical ratio, rather than a week per operation; the following ratio is more sound:

$$\text{Critical ratio} = \frac{\text{throughput time working hours}}{\text{total of operation times for jobs}}$$

This can range from 9 to 14 for the majority of companies—worse for the poorer ones.

Improvements in throughput should be instigated by means of better grouping, layout and facilities as indicated by Pareto analysis and production flow analysis.

CONTROLLING RELEASE OF JOBS

The more work is placed on the shop floor, the worse congestion becomes. Many companies now use input–output analysis to measure the flow of work into and out of different departments and base their release of jobs on this. The input and output of departments or work centres can generally be calculated by means of hours of work content.

TRACKING JOBS TO KEEP THEM ON SCHEDULE

Whether this is done centrally by the production control department, or is decentralized and done by sub-functions of production control (in larger shops) or by foremen/supervisors (in smaller shops where there is not too much complexity) will depend on the company operation. But it needs to be done, and the responsibility clearly defined.

MINIMIZING THE NUMBER OF EXCEPTIONS TO THE PLAN

One of the biggest problems most manufacturing units have to face is last-minute insertions or changes to priority of jobs instigated by the sales department or by top management; the insertion of one job causes all the others which follow to be rerouted and retimetabled. In some cases it is wise not to plan all the available capacity but to leave a percentage allowance against contingencies which arise.

With a good information system, alternative job routings through the shop by carefully planned schedules can lead to reduced congestion, less work in progress and more deliveries on time. Some computer systems have been introduced to assist with this information need.

17.5 Computerized scheduling systems

MAIN REQUIREMENTS

The main requirements for a computerized scheduling system are the following:

Work centre/calendar file

This is a matrix of the normal hours that are available for each work centre in the production unit.

A 'work centre' is defined as a group of machines or operators which are the same or very similar in terms of operating characteristics, or in terms of skills. No two machines or operators will be exactly alike (there will always be some variation) but provided they come within the specified tolerance range of options then they will be classed in the same work centre.

'Normal hours available' would be the usual hours normally available for that group of machines. A normal working day might be eight hours Monday to Thursday, seven hours on Friday. If machines or people are working in shifts their normal attendance or available hours are used. If overtime is normally worked each week then this should be included.

The calendar is usually allocated the normal hours available for each week that the factory is working. If there are holiday periods or normal stoppages for all the work centres, then no hours are allocated for these days.

Reference is normally by week number followed by day number, e.g., WWD (503 would refer to week 50, day 3, i.e., Wednesday).

Efficiency factor

An 'efficiency factor' is usually allocated for each work centre. This would be the utilization of the equipment or people in producing useful work. This factor is used in the loading and scheduling of jobs to the work centre.

Modifications file

This procedure is a temporary modification to the normal hours available, and is specific. A modification is usually made for reasons of the following types:

PRODUCTION SCHEDULING

(a) If additional hours are to be worked, e.g., 10 hours on work centre 322 on day 503.
(b) If breakdowns occur or if maintenance is needed, e.g., 0 hours on work centre 322 between 481 and 495.

The procedure is to increase or decrease the hours available for a specific period at a specific work centre.

Job file

Each order needs to be loaded into the computer records, with details such as:

- Customer name
- Order number
- Description of goods
- Product number
- Date of order
- Date due

Further details may also be added.

Component file

This file describes the parts, components or materials used on the job (parent). It will contain:

- Part number
- Drawing number (if needed)
- Part description
- Quantity per parent
- Unit of measure (each, metre, kg)

It will also include a network reference (e.g., 10–25) by means of which each part within a computerized scheduling system is linked to its parent.

Routing file

This file contains details of the operations to be performed on the part/component. It would normally contain, for each part number:

- Operation number
- Description
- Work centre reference
- Setting hours
- Operation hours per unit (each)

- Waiting hours (waiting after operation and inspection)
- Transport hours (time needed to transport work containers from one work centre to the next)

Note: Waiting and transport hours can be combined. Overlap can be accommodated if transport time is prefixed with a reference. Movement then takes place at h hours following the start of the operation, rather than h hours following the last operation.

The information from the above files is then processed. With most systems a priority index is calculated (see dispatching rules). In some instances multiple weighted indexes are used, in others simplified formulas, e.g.,

$$\frac{\text{Amount of process time to be completed} + A}{\text{total working hours to due date for order} + B}$$

where both $A + B$ are constants. By this, an index with a maximum of 99 is calculated, allowing for managers to override the calculation with their own priority index, which will be 100 plus.

Once the priority has been calculated, the computer then loads the facilities, starting on jobs with the highest priority. Queues of tasks are sequenced through the facilities and data calculated.

Each operation for each order is located in the system, with interlinks for the relationships between operations, and between operations and assembly.

1. *Setting time*: for preparing work or the machine for use.
2. *Operation time*: quantity × operation time per piece.
3. *Waiting time*: for inspection and administration or obtaining transport.
4. *Transportation time*: time to collect and carry to the next operation.

The computer then calculates, for each load centre, the total load in hours for each number. Typically, the first few weeks are overloaded, and as the weeks further ahead are considered so the load gets progressively smaller. This is often expressed in tabular and in 'histogram' form. This is planning with unlimited capacity.

Where necessary, the computer will also calculate the 'queuing time' for each operation, when, for example, more than one job is waiting to be started on a particular machine.

FORWARD AND BACKWARD SCHEDULING

This is the part of scheduling which needs the computer. Capacity is then fixed; no overloads are allowed, so the system has two options, as follows:

PRODUCTION SCHEDULING

1. *Fixed capacity—due date fixed* The computer works back from the due date using earliest start date (ESD), e.g., now, or with latest start date (LSD), where the operators are closed up as much as possible or can vary between the two options.
2. *Fixed capacity—start date fixed* Here the computer works from the start of the operations and ends up with a final date for completion.

Various options have then to be tried, such as overtime, offloading and consideration of priorities. The better the organization of capacity management, the more fruitful is scheduling likely to become.

COMPUTER OUTPUT REPORTS

The computer will normally produce the following three main types of screen-based information, which can also be printed out on documents.

MICROSS 82/1

Report 50   CURRENT WORK LOADS BY WORK CENTRE   Page 1

Work Centre CODE 1010
Work Centre Name . . . . . HYDRAPATH 3000
No. of UNITS 3
Performance Factor 81

	Available Capacity in hours			Load (in hours) based upon			
Week Nos.	NORMAL	PLANNED	CUMUL.	ESO	LSD	Cumul.	SSD
31	123	99	99	659	67	67	99
32	123	99	198	160	118	185	99
33	123	99	297	79	66	251	99
34	123	99	396	60	49	300	83
35	99	80	476	6	10	310	56
36	123	99	575	0	0	310	33
37–40	492	398	973	0	117	427	139
41–44	492	398	1371	0	122	549	56
45–58	492	398	1769	0	71	620	104
49–52	423	342	2111	0	62	682	85
1–4	492	398	2509	0	146	828	54
5–8	492	398	2907	0	79	907	44
9–12	492	398	3305	0	44	951	13
		SUB-TOTALS		964	951		964
		LATER		0	13		0
		TOTALS		964	964		964

Figure 17.5 Example of current work loads by work load centre

INDUSTRIAL ENGINEERING

Current work loads by work load centre

The example in Fig. 17.5 shows the loads based on finite-capacity scheduling, where the word 'overload' would indicate that queues are present at the facility.

Current work loads by work centre (*Fig. 17.6*)

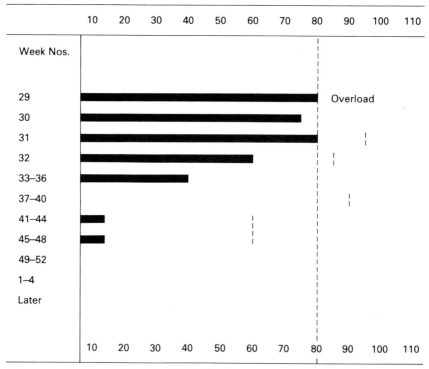

Figure 17.6 Example of current loads by work centre

This shows the data in more detail, in two main sections:

PRODUCTION SCHEDULING

1. Available capacity in hours: normal, planned, cumulative planned.
2. Load in hours, based upon:
   - ESD: earliest start date possible
   - LSD: latest start date possible
   - CUM: cumulative LSD
   - SSD: scheduled start date

The experienced manager can interpolate various facets of information about the facility so that he knows the load situation which exists for various types of schedule.

Work centre work to list

This is the list of tasks in order of importance as scheduled by the computer (Fig. 17.7). It shows the scheduled start date (SSD) and scheduled finished date (SFD) and previous and next work centres. The absence of a previous work centre indicates that the task is ready for the present operation.

COMPUTER SYSTEMS AVAILABLE

The following list gives details of the main computer systems used for shop

```
KEWILL SYSTEMS LIMITED             MICROSS 82 7                    07 JUL 82
Report 60      WORK CENTRE WORK TO LIST from 24.5.82 to 24.3.83    Page 1
                       Work Centre CODE........1000
                       Work Centre NAME.......RUMBLING
                       No. of UNITS.................1   Def A
                       Performance Factor......100

JOB      NETWORK                                    OP TIME  SSD   SFD   PREVIOUS   NEXT NEXT
NUMBER   REF      DRAWING No DESCRIPTION     QTY No hrs   ww.d  ww.d   W/C   W/C    W/C  SSD

22267    45-500   310423   STIFFENER         300   1   3.2  24.5  24.5                1003  25.1 C
22267    60-500   310425   TOP STIFFENER     150   1   1.3  24.5  24.5                1002  25.1 C
22725    0-0      153318   760m/8m CUTAWAY COUL 80 1   4.3  24.5  24.5                1002  25.2
22267   165-450   320018   BOSS              300   2   6.0  25.1  25.1         1001   1005  25.3 C
22267    35-500   310116   SUPPORT SKID      600   2   6.3  25.2  25.3         2011              C
22267   175-450   310122   GUIDE PLATE       150   3  47.7  26.1  27.2  1006   1004              C
22267   125-425   310122   GUIDE PLATE       150   3   1.3  27.2  27.2  1006   1004              C
22267    25-500   310419   RUNNER            450   3   6.3  27.3  27.3  1006   1004   1005 27.4 C
22781     0-0     242138   CLAMPING PLATE   2500   2  50.4  27.4  28.5*        1002     99  29.1 C
22751    70-80    320804   STOP              500   3   3.3  29.1  29.1  1006   1004
22751    50-80    320808   WRAPPER           500   4  11.2  29.4  29.5  1002   1003   1004 30.2
TEST JOB1 10-100 1234567   -AXTESTING LIBRARY COMP 10 50 14.9 30.3  30.4  2525   1004   1507 30.5
22759     0-0     241657            CENTRE MEMBER  1200 7 230.5 30.3  36.2  1001     99   1004 36.1
22732     0-0     242573   DEFLECTOR (LEFT HAND) 2450 4  20.4  40.1  40.3*  1004    407   2525 40.4 C
```

Figure 17.7 Example of work centre work to list

floor scheduling. Most packages available today will list scheduling but in fact they do not schedule properly—they merely load work centres from a requirements processing module, divide the lead time by the number of operations and then load those. Proper scheduling takes into account the delays caused by queuing—the major reasons for long throughput times through the shop.

MICROSS (*output reports previously illustrated*)

Designed and sold by Kewill Systems (address on p. 147). There are over 800 systems sold and they can be started on microcomputers, e.g., IBM PC, or AT, and upgraded to larger multi-screen, multi-tasking configurations. Kewill has many other options available which link to the shop scheduling module.

IBM CAPOSSE-E

IBM has had many years of experience of shop scheduling systems. This started as KRAS, named after the person who invented the original system. It was modified to become CLASS (Capacity Loading and Scheduling System) which had many uses during 1960s. This was then upgraded with enhancements to CAPOSSE-E. It is a module of the much larger IBM system, COPICS, and is used with the total CIM (computer integrated manufacturing) concept whereby CADCAM, manufacturing control systems and CNC/robotics/FMS are integrated.

WASP

This was developed by the United Kingdom Atomic Energy Authority (UKAEA), Harwell, and used in some of its supplying workshops. Some of the more modern systems have been developed using the basic principles of WASP. At present it uses punched-card input, but it is hoped that modified data entry will be available.

4W

This system was developed by the Production Engineering Research Association (PERA) at Melton Mowbray, Leicestershire. This is one of the more modern systems available, but at present cannot be effectively integrated with other production control modules. It is therefore used as a stand-alone system but can be interfaced to retrieve data firm serial files in other systems, e.g., Mapics, HDMS. It can be used with microcomputers (IBM XT) and

PRODUCTION SCHEDULING

minicomputing (ALTOS or PRIME). With this system there is a short-term schedule producing a work-to list for, say, five days. A finite-capacity planning module exists for both machines and operators, plus an infinite-capacity module.

Gantt charts

With the advent of graphics packages on microcomputers, some vendors will now provide Gantt charts which can be displayed on a monitor (or VDU). The main IBM and Olivetti agents would be the most likely source for this type of package. It is virtually a manual scheduling system represented by a microcomputer.

NEW APPROACHES TO SCHEDULING

Optimized Production Technology (OPT) provided by Creative Output (UK) Ltd, The Hounslow Centre, 1 Lampton Road, Hounslow, Middlesex (tel. 01-572 3111) has introduced a novel approach to some traditional problems—and broken many of the long-held rules about production planning and control.

The system was invented by an Israeli physicist named Eli Goldratt and introduced in the United States. The philosophy of the system is to identify potential bottlenecks in facilities. The flow of materials into these needs careful control: if facilities (machines or people) other than these bottlenecks are kept working at maximum utilization and efficiency they only build up inventory and work in process.

The OPT system utilizes a two-part package:

1. A simulated manufacturing program—used to identify 'bottlenecks'.
2. A set of shop floor management rules.

The following list sets out conventional rules against OPT rules.

What rule drives your business?

Conventional rules	*OPT rules*
– Balance capacity, then try to maintain flow.	1. Balance flow not capacity.
– Level of utilization of any worker is determined by its own potential.	2. Level of utilization is not determined by its own potential but by some other constraint in the system.

163

Conventional rules
- Utilization and activation of workers are the same.
- An hour lost at a bottleneck is just an hour lost at that resource.
- An hour saved at a non-bottleneck is an hour saved at that resource.
- Bottlenecks temporarily limit throughput but have little impact on inventories.
- Splitting and overlapping of batches should be discouraged.
- The process batch should be constant both in time and along its route.
- Schedules should be determined by sequentially:
  - Predetermining the batch size
  - Calculating lead time
  - Assigning priorities, setting schedules according to lead time
  - Adjusting the schedules according to apparent capacity constraints by repeating the above three steps

OPT rules
3. Utilization and activation of a resource are not synonymous.
4. An hour lost at a bottleneck is an hour lost for the total system.
5. An hour saved at a non-bottleneck is just a mirage.
6. Bottlenecks govern both throughput and inventories.
7. The transfer batch may not, and many times should not, be equal to the process batch.
8. The process batch should be variable not fixed.
9. Schedules should be established by looking at all of the constraints simultaneously. Lead times are the result of a schedule and cannot be predetermined.

17.6 Summary

There is no simple solution to the scheduling problem, as there is no one solution for every company. Each company should first of all identify its own problems, the likely causes of these, and then devise solutions. Any one of the concepts presented in this chapter could yield good results.

Like most techniques of management, successful use depends on a well-organized and thoroughly planned implementation, with the necessary training of personnel.

This chapter has introduced the main principles and concepts involved with scheduling systems in industry. Manual and computerized results suggest that improvements in output of 33 per cent are often accomplished.

APPENDIX
Summary of some common terms used

Job analysis The process of examining a job in detail in order to identify its component tasks so as to plan work measurement and method study projects. Information is collated by observation, interviews and questionnaires and by examinating existing records.

Job specification A product of job analysis—a detailed statement of the knowledge and physical and mental activities required to carry out the tasks which constitute the job. Based upon job analysis, but more detailed and considers the nature of the responsibilities according to a generalized scheme which is applicable to a range of jobs under broad headings such as mental strain, physical strain, conditions, etc., and also considers the qualifications, abilities and experience needed to discharge the responsibilities of the job accurately.

Method study Method study is the systematic recording and critical examination of existing and proposed ways of doing work as a means of developing and applying easier and more effective methods and reducing costs. Method study was developed from motion study and is the parent of a number of more specialized techniques such as value analysis, procedure study, work simplification, etc.

Work measurement The application of techniques designed to establish the time for a qualified worker to carry out a specified job at a defined level of performance. There are a number of different techniques that give different standards of accuracy and are applied to different types of work with differing degrees of job complexity.

Work study Generic term for those techniques, particularly method study and work measurement, which are used in the examination of human work in all its aspects and lead to systematic investigation of all factors which

Time study A work measurement technique for recording the times and rates of working for the elements of a specified job carried out under specified conditions, and for analysing the data so as to obtain the time necessary for carrying out the job at a defined level of performance.

Standard performance The rate of output which qualified workers will naturally achieve, without over-exertion, as an average over the working day or shift, provided they know and adhere to the specified method and provided they are motivated to apply themselves to their work. This performance is denoted as 100 on the British Standard Rating and Performance Scale.

Work content Defined as: basic time + relaxation allowance + any allowance for additional work, e.g., that part of a contingency allowance which represents work.

Standard time The total time in which a job should be completed at standard performance, i.e.: work content + contingency allowance + unoccupied time allowance + interference allowance, etc.

Contingency allowance A small allowance which may be added to the standard time to meet legitimate items of work or delay, the precise measurement of which is uneconomical because of its infrequent or irregular occurrence.

Basic time The time for carrying out an element of work at standard rating; it is calculated as follows:

$$\frac{\text{Observed time} \times \text{observed rating}}{\text{standard rating}}$$

Relaxation allowance An addition to the basic time intended to provide the worker with the opportunity of recovering from the physiological and psychological effects of carrying out specified work under specified conditions and to allow attention to personal needs.

Activity sampling A technique in which a large number of instantaneous observations are made, over a period of time, of a group of machines or workers. Each observation records what is happening at that instant and the percentage of observations recorded for a particular activity or delay is a measure of the percentage of time during which that activity or delay occurs. *Rated activity sampling* is an extension of this in which a rating is applied to each work element so that the work content may be calculated in addition to the percentage of time occupied by other activities or delays.

Synthesis A work measurement technique for building up the time for a job at a defined level of performance by totalling the element times obtained previously from time studies on other jobs containing elements concerned, or from synthetic data.

Synthetic data The name given to tables and formulae derived from the analysis of accumulated work measurement data, arranged in a way suitable for building up standard times by synthesis.

PMTS (predetermined motion time systems) Work study technique whereby times established for basic human motions (classified according to the nature of the motion and the conditions under which it is made) are used to build up a time for a job at a defined level of performance. There are many such systems in existence, some of which are: Work Factor, MTM (Methods-Time Measurement), Basic Motion Time Study, Motion Time Analysis (MTA).

Methods-Time Measurement (MTM) A procedure which analyses any method or operation into the basic motions required to perform it and assigns to each basic motion a predetermined time standard which is determined by the nature of the motion and the conditions under which it is made.

Gilbreth, Frank, 17

Hertzberg, Frederick, 19
Histogram, 102, 108, 109
Human factor, 27–30
   in industrial engineering, 29

Identity (ID) cards, 142
Incentives, 89
Industrial engineering:
   common terms used, 166–168
   data collection, 140, 141
   the key, 3–9
Industrial engineers, 1, 3–5
   personal attributes, 30
International Labour Office (ILO), 58
Inventory, 130

Job, 148
Job file, 157
Job tickets, 140
Just in time (JIT), 40

Labour productivity, 5
Latest start date (LSD), 161
Leadership, 25, 26
Line balancing, 45
Logic controllers, 53
Logistics charts, 39
Lost time, 141

Magnetic character recognition (MCR), 143
Maintenance work standards, 86–89
Management, 21–26
   historical developments, 16–20
   introduction, 21
   scientific, 16–18, 22
   social and psychological aspects, 18–20, 22–24
   systems, 24
Management skills, developing, 25, 26
Managerial engineering, 40–51
Maslow, Abraham, 18
Mathematical models, 34
Maxi MOST, 71
Maynard, H.B. & Co. Ltd, 71, 83
Maynard Operation Sequence Technique (MOST), 46, 50, 51, 60, 62–74

McGregor, D., 18
Mean, 105, 111
Mean absolute deviation (MAD), 107
Measured daywork, 89
Median, 105, 106
Method study, 12, 13, 166
Methods engineering, 39–51
Methods index, 6
Methods time measurement (MTM), 17, 50, 51, 57–60, 62, 65–67, 75, 84, 108
Microelectronics, 53
MICROSS, 162
Mini MOST, 71
Mode, 106
Modelling, 33, 34
Monitoring, 144, 147
Motion economy, 49
Motivation, 28
MRP II, 146
Multiple activity chart, 36
Multiple regression analysis (MRA), 14, 85

Normal distribution, 114–120

Observation, 96
Ogives, 104, 110
Operational research (OR), 128–135
   definitions and activities, 129–134
   introduction, 128, 129
   methodology, 134, 135
Operations, 1
Optical character recognition (OCR), 143
Optimized Production Technology (OPT), 163, 164
Organization and methods (O & M), 84
Outline chart, 36
Output records, 141
Overall productivity, 5

Parametric representation, 105
Performance, 6, 7
Personal needs, 61
Piecework, 89
Population measures, 121
Practices, 154–156
Predetermined motion time systems (PMTS), 62–74
   third generation techniques, 62, 63

Index

Action-Centred Leadership, 26
Adair, J., 26
Added value, 89
Allocation, 130
Assembly, 54
Attendance time, 141
Authority, 25, 26
Automation, 52–54
   applications, 53
   low cost, 53, 54

Bar-codes, 143
Basic time, 61
Behavioural management, 18, 22
Bench marking, 85, 86
Brainstorming, 49
Breakdown, 141
British Standard (BS), 50, 58

CADD, 47
CAPES, 51, 78, 80–82
CAPOSSE-E, 162
Cause and effect diagram, 40, 41
CEPS, 75–83
Chart, process, 34
Charting symbols, 35
Clerical MOST, 74
Clerical work measurement (CWM), 83–85
Clock card, 140
Coefficient of variation, 108
Competition, 133
Computerized systems:
   applications in work measurement, 75–77
   estimating and planning, 75–91
   equipment monitoring devices, 144–146

   equipment suppliers 146, 147
   for data collection, 141
   input forms, 142, 143
   output forms, 143, 144
   production scheduling, 156–164
   time study, 77–83
Contingency, 61, 65
Creative thinking, 49
Critical ratio, 155
Cumulative frequency, 101

Data capture for computers, 141–147
Data collection, 139, 140
Daywork performance, 58
Descriptive statistics, 97
Diploma of the Institute of Management Services, 11
Dispatching, 148, 149

Earliest start date (ESD), 159, 161
Estimating and confidence intervals, 121–127

4W, 162
Facility layout, 41–46
   and design, 41
   fixed position layout, 46
   procedure, 46–49
   process layout, 45–46
   product layout, 44–45
Fatigue, 61
Financial models, 34
Flexible manufacturing systems (FMS), 44
Flow process chart, 36
Frequency distribution, 99

Gantt chart, 148, 149, 163

169

INDEX

Prediction, 96
Preventative maintenance (PM), 88, 89
Priority, 148
Process charting, 34–39
  logistics charts, 39
  proximity charts, 39
  recording charts, 35–37
  string diagrams, 37–39
Process layout, 45
Production, fallacies, 49–51
Production control packages, 162, 163
Production scheduling, 148–165
  computerized systems, 156–164
  definitions, 148, 149
  defining service levels, 151–154
  dispatching rules, 149–151
Productivity, 5–9
  centres, 7
  improvement, 8, 9
  measures, 5–7

Quality of work life, 7
Questioning technique, 47, 48
Queuing, 131

Random sampling, 97
Range, 107
Relaxation allowances, 61
Replacement, 132
Robot time and motion, 83
Route card, 140
Routing, 132
Routing file, 157

Sampling, 97–99
  error, 98
Scientific management, 16–18
Scientific method, 16, 22
Sequence controllers, 53
Sequencing, 132
Service, 148
Service levels, 151
Setup reduction, 40, 42, 43
Shop floor, 137–165
Simulation, 34
Simultaneous-motion charting ('Simo' chart), 37
Smith, A., 16
Software packages, 76, 78–85
Standard deviation, 107, 111, 112
Standard hour (SH), 13, 59, 61

Standard minute (SM), 13, 59, 61
Standard performance, 59
Standard time, 59, 61
Statistical methods:
  central tendency measures, 105–113
  collection data, 97–99
  describing data, 99–105
  dispersion measures, 107, 108
  introduction, 95, 96
Statistics, 95, 100
Systematic sampling, 98
Systems management, 24

Task, 148
Tavistock Institute of Human Relations, 20
Taylor, F., 20
Terminals, 141
Theory of added variance, 64
Time and attendance, 146, 147
Time slotting, 85
Time standards, 13, 58
  purpose of, 56–60
Training, 27, 28
Travel chart, 30

Universal maintenance standards (UMS), 87
Utilization, 6, 153

Variance, 107, 122
Verification, 96
Voice recognition, 144
Vroom, V. H., 19

WASP, 162
Work centre, 156
Work measurement, 12–14, 57–74, 166
  computerized systems, 75–91
  determining work content, 63–67
  and incentives, 89–91
  introduction, 57, 58
  techniques, 14, 60, 61, 85, 86
Work measurement techniques, 14, 60, 61, 85, 86
Work standards for maintenance, 86–89
Work study, 10–20, 166
  definitions, 12, 13
  introduction, 10–15
  need for, 11, 12
Work to list, 161
Worker, 27

Lanchester Library